GALIT SHMUELI

PRACTICAL ACCEPTANCE SAMPLING

AXELROD SCHNALL PUBLISHERS

Contents

To my husband Boaz who made the production of this book a reality

Preface

The purpose of this textbook is to introduce the reader to acceptance sampling, also called sampling inspection. While the focus of textbooks has shifted from acceptance sampling to process monitoring and control charts, acceptance sampling is still widely applied across a wide range of industries. The use of ISO or ANSI/ASQC acceptance sampling standards is common and frequently a requirement, and therefore many manufacturers, suppliers, and purchasers employ such methods. In higher education, acceptance sampling is often taught as a module in a course on quality control or industrial statistics. In other cases, it is offered as a separate course in a quality control program. This book is aimed at providing a readable and practical textbook in such courses. It is also a useful introduction for professionals who strive to better understand acceptance sampling. From our experience, learning is best achieved by doing. Hence, we designed the book to achieve self-learning in the following ways:

- The book is relatively short compared to other books on acceptance sampling, to reduce reading time and increase hands-on time.

- Explanations strive to be clear and straightforward with more emphasis on concepts than on theory.

- Each chapter has end-of-chapter problems, ranging in focus, with several requiring computations using basic software such as Excel or online calculators.

The book is designed for a 5-6 week module that is suitable for graduate or upper-undergraduate level. A pre-requisite is a basic course in statistics and probability that covers distributions such as the Binomial, Hypergeometric and normal distributions, the concept of expected values, and summary statistics. A suggested schedule is:

Weeks 1-2   Chapters 1 and 2 (Introduction, terminology, and single-stage plans for attributes). It is advisable to include a refresher on the Binomial, Hypergeometric and normal distributions.

Week 3   Chapters 3-4 (double-stage sampling plans, rectifying inspection plans).

Week 4   Chapter 5 (Plans for variables (measurement data)).

Week 5   Chapter 6 (Continuous sampling plans)

Week 6   Project presentations or final exam.
   We strongly recommend including a team project in such a course, where students work on a real problem where acceptance sampling can be beneficial.

   In terms of software, Microsoft Excel and online calculators are used throughout the book to illustrate the different methods and procedures. The reason for choosing Excel is due to its wide accessibility in industry and the prior familiarity of most students with it. We use functions in Microsoft Excel 2010 which are also available in earlier versions of Excel (2003 and 2007), although function names might be slightly different. We also use Excel for creating charts, although careful manual formatting is necessary for producing effective graphs.

What's new in the second edition?

The second edition provides several improvements, based on feedback from readers and on software updates. The book now includes a section on Accept-on-Zero plans. Additional screenshots from the newly-designed SQCOnline.com illustrate several new calculators. Finally, the second edition offers an improved design for enhanced readability.

Supporting materials for this book are available at `www.galitshmueli.com/practical-acceptance-sampling`.

1 *Introduction*

1.1 *Acceptance Sampling in Industry and Commerce*

In today's manufacturing world, the path from raw materials to final product often takes place over multiple companies and across multiple continents. A company selling laptop computers has likely sub-contracted different parts to different sub-contractors. The assembly might take place in yet another plant. And the packing and shipping, from yet another location. In order to assure a certain quality level, companies use inspection at the different supply chain stages.

   Acceptance Sampling, also known as *Sampling Inspection*, consists of quality assurance schemes designed to test whether the quality of batches of products or services conform with requirements, based on inspecting *only a sample* from each batch. The use of sampling inspection relies on the premise that products need not conform 100% with specification requirements, and it is often more economical to allow a small percentage of non-conforming items to pass on for later rejection than to bear the expense of 100% inspection. Acceptance sampling provides criteria and decision rules for determining whether to accept or reject a batch based on a sample. Civilian ISO acceptance sampling plans and their military counterparts (Mil-Std) are commonly used standards in industry. These plans dictate the sample size to be drawn from each batch, and the requirements that the sample must meet to assure that the entire batch is of acceptable quality.

   Acceptance sampling is useful for testing the quality of batches of items, especially when a large number of items must be processed in a short time. However, acceptance sampling is not a method for monitoring or improving the quality of a process. It is important to keep this in mind and to use other quality control tools such as control charts for monitoring and designed experiments for improvement.

The origin of Acceptance Sampling is in World War II, for testing the quality of bullets in the United States' military. The need for sampling arose from the destructible nature of bullet testing. In general, acceptance sampling is needed either when inspection is destructive, or when inspecting each item (100% inspection) is prohibitively expensive or time-consuming.

Acceptance sampling is commonly used in contracting and sub-contracting, where the contractor wants to assure the quality of the incoming goods or services. For example, a school providing lunch services inspects the food quality on a daily basis. A large cosmetics company inspects the quality of the products in each shipment that it receives from its overseas manufacturers. A large mining manufacturing company in the USA uses acceptance sampling to inspect machined parts from its vendors. The famous coffee chain Starbucks uses acceptance sampling to assure the quality of coffee shipments from its different suppliers.

> Because of Starbucks' high volume, coffee exporters competed to become suppliers. For its part, Starbucks cultivated long-term relationships with its suppliers, providing training working closely with them. For quality control, Starbucks extracted three different samples from every shipment: one before the export was arranged, one just before shipment and one on arrival at the roasting plant. At each stage Starbucks reserved the right to refuse the shipment[1].

[1] from *The Starbucks Brand* case study, Rotman School of Management, University of Toronto, www.rotman.utoronto.ca/bic/caseseries/PDFs/starbucks.pdf, accessed Aug 1, 2011.

Acceptance sampling is also commonly used by manufacturers or suppliers themselves, as a tool for audit or compliance. A well-known vendor of child safety car seats inspects the quality of its products using acceptance sampling. A call center audits a sample of calls in each shift to assure quality service. A manufacturer and exporter of home textiles in India employs inspection after the fabric cutting stage, after sewing, and after packing. Manufacturers of electronic parts in China use sampling inspection to assure that shipments of their parts will be up to the standard of their purchasing companies.

1.2   Who Should Understand Acceptance Sampling?

Anryone involved in the inspection phase should be literate of what acceptance sampling is, how to deploy it effectively, and what sampling plans guarantee. This includes management, R&D, and quality personnel. In particular, production and process engineers, manufacturing and design engineers, purchasing agents and managers, and quality engineers, inspectors, supervisors, and managers.

1.3   What Are Sampling Plans?

Acceptance Sampling is based on the notion of *probability*: Although we cannot deduce from a sample about the exact quality of the entire batch, we can reach conclusions with a given certainty level. For example, using a sampling plan and examining a sample, we might conclude that "we are 95% certain that the percent of non-conforming items in the entire batch is no more than 1%".

Sampling plans consist of *criteria*, such as the size of the sample to inspect and *decision rules*, which tell us whether to accept or reject the batch, based on the quality of the inspected sample.

Samples are typically *drawn at random*. Drawing a random sample is sometimes operationally costly or inconvenient, for instance, when a shipment consists of many boxes that are piled one on top of the other. However, random sampling is extremely important for drawing correct decisions. Although it might be easier to sample all items from the top-most box, the sample information might be biased if the top-most box is slightly different in quality from all the other boxes in the shipment.

Sampling plans are divided into two general categories, depending on the quality property that is being measured:

Sampling plans for attributes are used for pass/fail (go/no-go) classifications such as whether a phone call disconnects, and for the number of non-conformities, such as the number of disconnected calls per hour.

Sampling plans for variables are used for continuous measurements such as the length of a phone call. Sampling plans can also range from single-stage plans, where a single sample is drawn from a batch, to double-stage plans, where one or two samples are taken, to multiple-stage plans, where one or more samples are drawn.

1.4 Terminology

Lot, batch, and shipment are used interchangeably to denote the collection of items to which an accept/reject decision must be made. The size of a batch is denoted by N.

Sample refers to a subset of a batch that is typically drawn at random, and is fully inspected. Sample size is denoted by n.

Producer refers to the party who manufactures or creates the good or service of interest. The producer is assumed to produce items on a regular basis.

Consumer refers to the party who purchases goods or services from the producer. The consumer receives batches of goods or services from the producer.

We use X to denote the random variable measuring the quantity of interest in the sample. For pass/fail data, X measures the number of failures in the sample. For example, if a sample of phone calls contains 3 disconnected calls, we denote this by $X=3$. For continuous measurements we use X_i to denote the measurement of the ith item in the sample. For example, if the first milk carton in a sample of 10 cartons contains 500cc, we use the notation $X_1=500$.

p is used to denote the proportion of non-conforming ("failed") items in the batch.

1.5   What determines a sampling plan?

Producer's Side

A producer produces items on a continuous basis (think of a production line). Let us assume that the production line quality meets a baseline requirement called *Acceptable Quality Level* (AQL). The producer would like to assure that batches from this process will be accepted with a high probability.

The producer's risk (denoted α) is the chance that a batch from a process with quality AQL is rejected.

Consumer's Side

The consumer cares about the quality of individual batches. The consumer's required per-batch quality is called *Lot Tolerance Percent Defective* (LTPD). The consumer would like to reject batches with quality LTPD or worse with high probability.

The consumer's risk (denoted β) is the chance that a batch with quality LTPD or worse is accepted.

A *sampling plan* is a scheme aimed at balancing the requirements of the producer and the consumer. It does so by *balancing the producer's risk (α) and the consumer's risk (β)*.

We will formalize these concepts in the next sections, within the context of specific sampling plans.

1.6   Problems

1. What are two advantages of acceptance sampling over 100% inspection?

2. What is the main purpose of acceptance sampling?

3. *Bee Healthy* is a coop of honey collectors in India. The coop regularly inspects batches of their honey jars to assure that quality is up to the standards of an organic certifying agency. A honey jar can either be conforming or non-conforming. Each batch consists of 300 jars.

(a) Should *Bee Healthy* use sampling plans for attributes or for variables? Explain.

(b) Is *Bee Healthy* considered the producer or the consumer? Explain.

(c) What is the meaning of the producer's risk in this case?

(d) How should a sample be drawn from each batch of 300 jars?

2   Single-Stage Inspection Plans for Attributes (Categorical Measurements)

A single-stage sampling plan for attributes means that a random sample of size n is drawn from the batch of size N. Then, the number of non-conforming items in the sample is counted. If this number exceeds a limit c, then the entire batch is rejected. Otherwise, it is accepted. c is called the *acceptance number*.

Hence, a single-stage sampling plan for attributes has two parameters: sample size (n) and acceptance number (c). In some cases, there will also be a third parameter called the *rejection number* (r) – we will discuss this in Section 2.4.

2.1   Computing Acceptance Probabilities

To understand how a sampling plan (sample size and acceptance criterion) is designed, we first introduce common notation and terminology. Consider a sampling plan with sample size n, such that n items are to be drawn at random from a batch of N items. The acceptance criterion is to accept the entire batch if there are c or less non-conforming items.

Let X denote the random variable counting the number of non-conforming items found in the sample. X can obtain values between 0 (all items conforming) and n (all items non-conforming). The batch is accepted if $X \leq c$.

The *probability of accepting the batch*, given that it has non-conforming proportion p, is called the *Operating Characteristic*

(OC). We can write this probability as:

$$OC(p) = P(X \le c|p). \qquad (2.1)$$

The notation | means "given". We therefore read equation (2.1) as:

> The probability of accepting a batch with quality p is equal to the probability of obtaining c or less non-conforming items in the sample, given that the batch quality is p.

To compute probabilities, let us consider a numerical example.

Example: Acceptance Sampling of Automobile Tires

Consider a large retailer of automobile tires in Canada who receives shipments from its supplier in Mexico. The typical shipment size is N=1,000 tires. A tire is considered non-conforming if it does not pass a series of safety tests. The production line in Mexico produces tires with an average quality level of AQL=1%. The retailer would like to assure that the percent of non-conforming tires per batch is not higher than LTPD=2

Consider a sampling plan where the retailer draws 100 tires and inspects them for safety. If there are 2 or less non-conforming tires, then the entire batch is accepted. Otherwise, it is rejected. This sampling plan has parameters n=100 and c=2.

Consumer's View: Per-Batch Computations

From the consumer's perspective, the focus is on a particular batch. Therefore, the probability of accepting the batch depends on the batch size, on the sample size, and on the acceptance threshold.

The probability of accepting a batch of tires, where the true proportion of non-conforming tires is 0.01 (or 10 tires) can be written as:

$$OC(0.01) = P(X \le 2|p = 0.01, n = 100, N = 1,000). \qquad (2.2)$$

Computing this probability depends on the distribution of the random variable X, which counts the number of non-conforming tires in a sample of 100 tires drawn from a batch of 1,000 tires with 1% non-conforming tires. The distribution that describes this scenario is the Hypergeometric distribution. A Hypergeometric random variable counts the number of "successes" in a sample of size n, which is drawn at random from a population of size N. The Hypergeometric probability in general can be written as:

$$OC(p) = P(X \leq c|p, n, N) = \sum_{x=0}^{c} \frac{\binom{Np}{x}\binom{N-Np}{n-x}}{\binom{N}{n}}. \qquad (2.3)$$

In our example:

$$P(X \leq 2|p = 0.01, n = 100, N = 1,000) = \sum_{x=0}^{2} \frac{\binom{10}{x}\binom{1,000-10}{100-x}}{\binom{1,000}{100}}.$$

This quantity can be easily computed in software such as Microsoft Excel, Minitab, or online probability calculators. In Excel 2010, use the function =$HYPGEOM.DIST$ (Note that the number of "successes" in the population is $0.01 \times 1,000 = 10$). This is shown in Figure 2.1. Or, we can use online calculators such as the Hypergeometric calculator[1] on www.stattrek.com (see Figure 2.2). In both cases, we obtain the result $OC(0.01)=0.9308$. This means that if this sampling plan is used, then a batch with 1% non-conforming tires will be accepted with probability 0.9308.

[1] http://stattrek.com/Tables/Hypergeometric.aspx

Producer's View: Manufacturing Process Computations

The producer is not interested in specific batches but rather in the ongoing manufacturing process. Computations of interest to the producer are therefore the probability of batch acceptance given an infinite population size. The probability of accepting a batch of size N given a sampling plan with parameters n, c is therefore computed using the Binomial distribution. The Binomial random variable counts the number of successes in a sample of n trials, where the probability of success for each item is given by p.

We can write the probability of acceptance as:

$$OC(p) = P(X \leq c|p, n) = \sum_{x=0}^{c} \binom{n}{x} p^x (1-p)^{n-x}. \qquad (2.4)$$

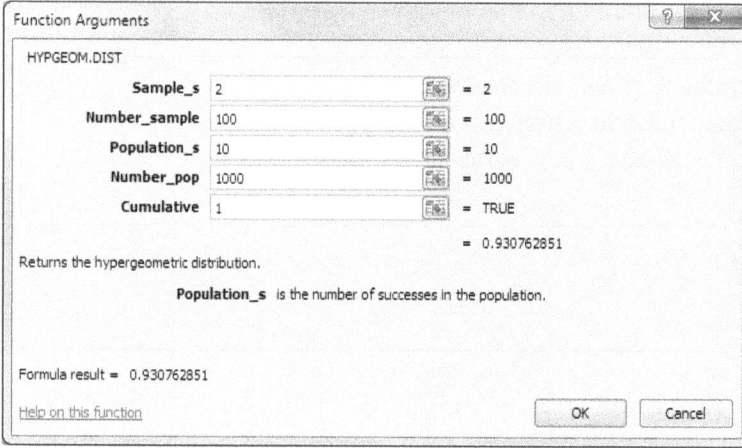

Figure 2.1: Computing acceptance probabilities using Excel 2010's HYPGEOM.DIST function

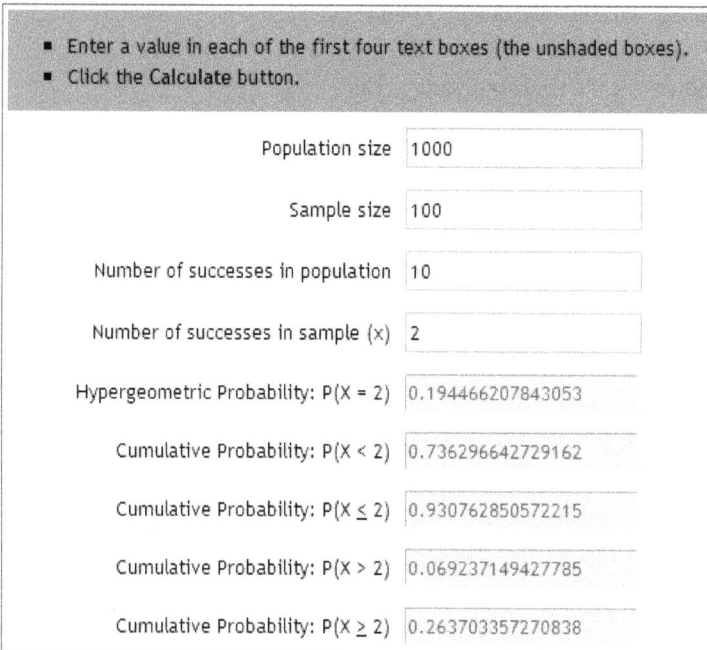

Figure 2.2: Using an online calculator (stattrek.com) to compute acceptance probability with the Hypergeometric distribution

For our example, the probability of acceptance, given a production quality level of AQL=0.01, is

$$OC(0.01) = P(X \le 2 | p = 0.01, n = 100) = \sum_{x=0}^{2} \binom{100}{x} 0.01^x 0.99^{100-x}.$$

Here too we can use software such as Excel or online probability calculators to do the computations. In Excel 2010, we use the function =*BINOM.DIST* as shown in Figure 2.3, or we can use www.stattrek.com's online Binomial calculator (see Figure 2.4)

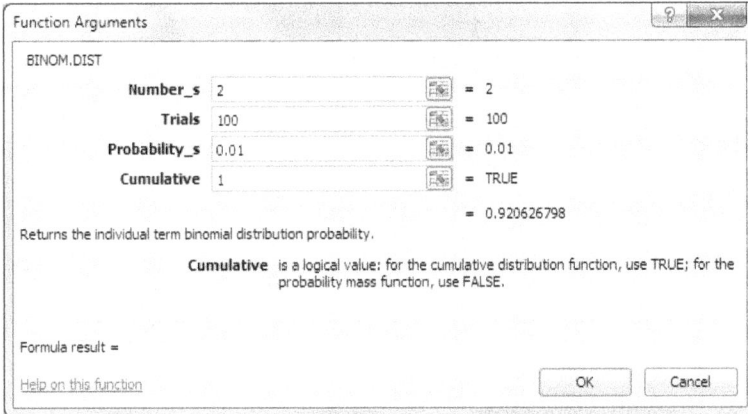

Figure 2.3: Computing acceptance probabilities using Excel 2010's BINOM.DIST function

Equivalence of Producer and Consumer Computations in Large Batches

The producer computation of batch acceptance is done based on an infinite production, whereas the consumer computation is based on a single batch. The two computations can yield slightly different probabilities, as we can see in our example: the probability of acceptance with $N=1,000$, $n=100$, $c=2$, and $p=0.01$ is 0.93 using the consumer batch-focused computation and 0.92 using the producer's production line-focused computation. When the batch size is more than 10 times the sample size, the two types of calculations give practically identical results.

In general, computations in standards such as ISO and military standards follow the producer's point of view. The logic

- Enter a value in each of the first three text boxes (the unshaded boxes).
- Click the **Calculate** button.
- The Calculator will compute Binomial and Cumulative Probabilities.

Probability of success on a single trial	0.01
Number of trials	100
Number of successes (x)	2
Binomial Probability: P(X = 2)	0.184864818824863
Cumulative Probability: P(X < 2)	0.735761978922956
Cumulative Probability: P(X ≤ 2)	0.920626797747819
Cumulative Probability: P(X > 2)	0.079373202252181
Cumulative Probability: P(X ≥ 2)	0.264238021077044

Figure 2.4: Using an online calculator (stattrek.com) to compute acceptance probability with the Binomial distribution

is that the relationship between the consumer and producer is expected to be a long-term relationship rather than a single batch transaction. For this reason, from now on we use only production-line type acceptance probabilities, using the Binomial distribution.

Let us compute the producer and consumer's risks for this plan, recalling that AQL=1% and LTPD=2%: The producer's risk is given by:

$$
\begin{aligned}
\alpha \;=\; & \text{Producer's risk} = P(\text{reject batch}|p = AQL) \\
=\; & P(X > 2|p = AQL) = 1 - P(X \le 2|p = 0.01) = 1 - 0.9206 = 0.0793.
\end{aligned}
$$

The consumer's risk is given by:

$$
\begin{aligned}
\beta \;=\; & \text{Consumer's risk} = P(\text{accept batch}|p = LTPD) \\
=\; & P(X \le 2|p = LTPD) = P(X \le 2|p = 0.02) = 0.6767.
\end{aligned}
$$

2.2   The Operating Characteristic (OC) Curve

We have seen that a single-stage sampling plan includes two parameters: the sample size (n) and the acceptance number (c). We can compute the probability of accepting a batch of any quality using the formula for $OC(p)$ in equation (2.1). A popular and useful plot is the Operating Characteristic (OC) Curve, which shows for each quality level (p) the acceptance probability ($OC(p)$).

The purpose of OC curves is to show the discriminating power of a sampling plan and to compare sampling plans. An ideal sampling plan completely discriminates between "good" and "bad" batches. Let us consider the producer's context (where the focus is on ongoing production). An ideal OC curve has $OC(p) = 1$ for $p \leq AQL$ and $OC(p) = 0$ for $p > AQL$, as shown in Figure 2.2. Such a curve would only be the result of employing 100% inspection, such that we would know the exact AQL. In reality, the steeper the OC curve around the value $p = AQL$, the more discriminating the sampling plan. Note that we can also devise batch-focused OC curves that use LTPD as the acceptable per-batch quality level.

An OC curve can be easily computed and plotted for a given sampling plan using software such as Excel. For example, for the sampling plan that we described for the tire example, we can use Excel to compute $OC(p)$ for various values of p and create an OC curve as follows (as shown in Figure 2.6):

1. Create a column p of values starting with 0 and increasing in small steps (e.g., 0, .01, .02,...)

2. In the next column, create a column for $OC(p)$ using the formula in equation (2.1). Replace the "number of successes in the population" with $N \times p$ (where p is the value in the left cell, and N is the batch size)

3. Insert a line chart such that the column for p is on the x-axis and the column for $OC(p)$ is on the y-axis. Make sure to format the chart to remove redundant gridlines, legends, and the y-axis range should be limited to 0-1.

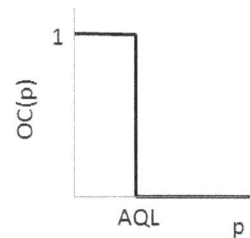

Figure 2.5: Ideal OC curve (producer focused)

Note that the consumer's risk is simply a point on the OC curve! It is equal to $OC(p = LTPD)$. This point is marked in Figure 2.6 as a square. Similarly, we can see 1-producer's risk as another point on the OC curve (marked with a circle).

p	OC(p)
0	1
0.01	0.920627
0.02	0.676686
0.03	0.419775
0.04	0.232143
0.05	0.118263
0.06	0.056613
0.07	0.025789
0.08	0.011273
0.09	0.004756
0.1	0.001945
0.11	0.000773
0.12	0.000299
0.13	0.000113
0.14	4.18E-05
0.15	1.51E-05
0.16	5.35E-06
0.17	1.85E-06
0.18	6.29E-07
0.19	2.09E-07
0.2	6.83E-08

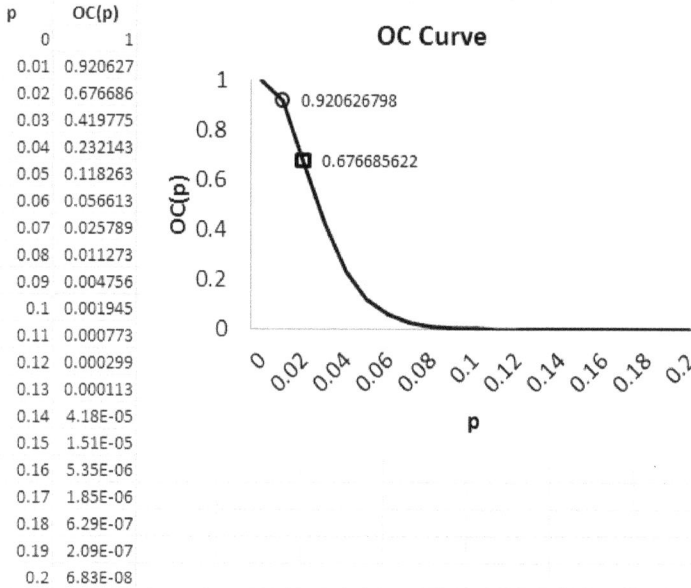

OC Curve

0.920626798

0.676685622

Figure 2.6: Using Excel to compute and plot the OC curve for the sampling plan in the Tire Example

Effect of n and c on Acceptance Probabilities and Risks

How does increasing the sample size affect the probability of acceptance? How does it affect the producer and consumer risks? The meaning of increasing the sample size is that we have more information. More information means less uncertainty, and hence lowered risks. Therefore, increasing the sample size will reduce both producer and consumer risks.

Increasing the sample size helps us better discriminate between low and high quality batches. Hence, the OC curve for a larger sample will have a steeper slope: higher acceptance probabilities for low values of p (=high quality) and lower acceptance probabilities for high values of p (=low quality). This is illustrated in Figure 2.7.

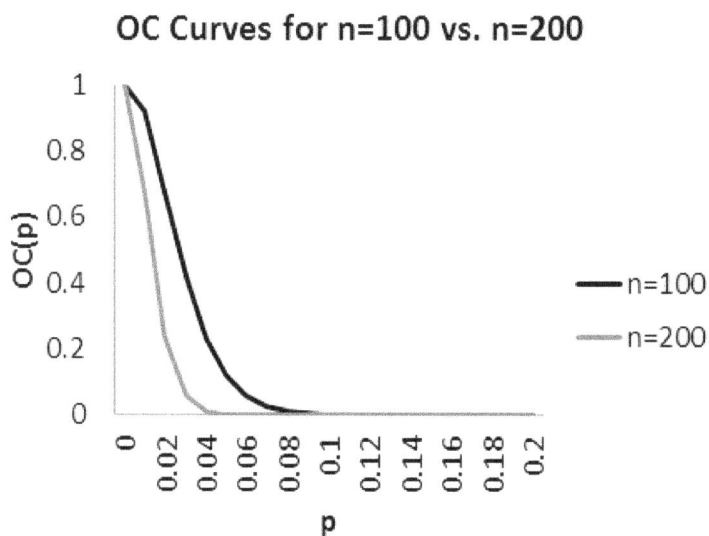

Figure 2.7: Effect of increasing sample size on the OC curve

How does changing the acceptance number affect the probability of acceptance and the producer and consumer risks? In general, decreasing c will make it more difficult to accept any batch, thereby decreasing the chance of accepting a good batch (1-producer's risk) as well as decreasing the chance of accepting a bad batch (consumer's risk).

In the sampling plan that we used earlier (n=100, c=2), the consumer's risk (of accepting a low-quality batch) is quite high (0.6767), while the producer's risk is very low (0.0793). To get a more balanced pair of risks, we could decrease the acceptance threshold (c). To see the effect of decreasing the acceptance number, in Figure 2.8 we overlaid the OC curve for the original sampling plan (n=100, c=2) with the sampling plan that has parameters n=100, c=3.

2.3   Designing a Sampling Plan

There are a few different approaches for designing sampling plans: One approach controls the producer's risk (α) for a given

OC Curves for c=2 vs. c=1

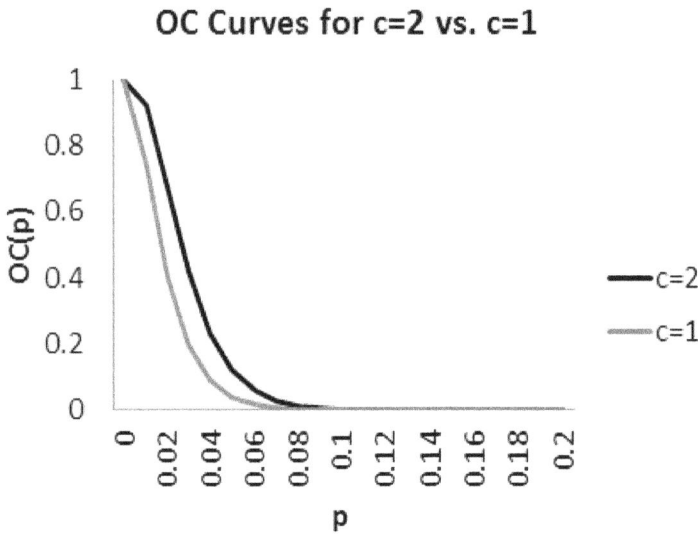

Figure 2.8: Decreasing the acceptance number c leads to lower acceptance probabilities

AQL. Another approach controls the consumer's risk (β) for a given LTPD. A third approach tries to control both the producer and consumer risks for given AQL and LTPD levels.

Finding a pair (n, c) in each of these approaches involves solving equations involving Binomial or Hypergeometric probabilities. For ease of use, tables or approximations have been devised using such calculations. The most popular plans used in industry for acceptance sampling are the ones based on controlling the producer's risk, which is the approach behind the military and civilian standards described in Section 2.4. In the next section we describe the ANSI/ASQC Z1.4 tables and a web calculator that simplifies the process of obtaining a sampling plan. In Section 2.6 we describe an approach for obtaining a plan that controls both producer and consumer risks for given AQL and LTPD. An approach that controls the consumer's risk for a given LTPD is described in Chapter 4, in the context of rectifying inspection.

2.4 ANSI/ASQC Z1.4 (Military Standard 105-E) Tables

The most widely used sampling plans are those based on the Military Standards, which originated in World War II. The military standard for attribute data is called Mil-Std-105, with the last version being Mil-Std-105E (issued in 1989 and finally cancelled in 1991). The military standards were then revised by a team of American, Canadian and British experts to create a standard for all three countries. The equivalent civilian counterpart by the International Standardization Organization (ISO) is called *ISO 2859*. The equivalent standard by the American National Standard Institute (ANSI) is *ANSI/ASQC Z1.4*. The equivalent British standard is *BS 6001*.

Many companies and organizations use the ISO 2859 (or equivalent) standard because it is required for certification. For example, the World Health Organization (WHO) guidelines require pharmaceuticals to use ISO 2859 sampling plans for inspecting packaging materials and finished products.

Design of Sampling Tables

The ISO (and equivalent) tables give inspection plans for a given batch size (N) and Acceptable Quality Level (AQL). These plans are therefore producer-based, assuming a long-term relationship between the producer and consumer. The plans guarantee a producer's risk (α) between 1% and 10%. To control the consumer's risk, there are various inspection levels.

Table Usage

The ANSI/ASQC Z1.4 and ISO 2859 standards are available for purchase in hardcopy from various organizations. The tables can be downloaded in PDF format from `www.sqconline.com/download`.

Code Letter

Finding a sampling plan starts with specifying the *batch size* (lot size) and the *inspection level*. Figure 2.9 shows the table where the lot size and inspection level determine a code letter. The default inspection level is level II. Inspection level I leads to more lenient inspection (which means higher consumer risk), and inspection level III leads to more stringent inspection. The Special Inspection Levels (S-1 to S-4) are used when small samples are required (for example, when inspection is prohibitively costly).

In our tire example, for level II inspection with lot size $N=1000$ the code letter is J.

Table I.   Sample-Size Code Letters.

Lot or batch size			Special Inspection Levels				General Inspection Levels		
			S-1	S-2	S-3	S-4	I	II	III
2	to	8	A	A	A	A	A	A	B
9	to	15	A	A	A	A	A	B	C
16	to	25	A	A	B	B	B	C	D
26	to	50	A	B	B	C	C	D	E
51	to	90	B	B	C	C	C	E	F
91	to	150	B	B	C	D	D	F	G
151	to	280	B	C	D	E	E	G	H
281	to	500	B	C	D	E	F	H	J
501	to	1,200	C	C	E	F	G	J	K
1,201	to	3,200	C	D	E	G	H	K	L
3,201	to	10,000	C	D	F	G	J	L	M
10,001	to	35,000	C	D	F	H	K	M	N
35,001	to	150,000	D	E	G	J	L	N	P
150,001	to	500,000	D	E	G	J	M	P	Q
500,001	and	over	D	E	H	K	N	Q	R

Figure 2.9: Code letter table. The first step is to determine the code letter based on batch size and required inspection level

Inspection Types

The next step is to choose between three types of inspection: normal, tightened, and reduced inspection. The default is normal inspection. At the beginning of inspection, normal inspection is used. The types of inspection differ as follows:

Tightened inspection (for a history of low quality) requires a
  larger sample size than normal inspection.

Reduced sampling (for a history of high quality) has a higher
  acceptance number relative to normal inspection (hence, easier
  to accept the batch).

There are special *switching rules* between the three types
of inspection, as well as a rule for discontinuation of inspection. These rules are empirically based. A schematic of the
rules is shown in Figure 2.10. It is important to note that the
ANSI/ASQC Z1.4 standard requires using the switching rules,
because we are assuming a long-term relationship between the
producer and consumer. Ignoring the switching rules will lead to
unexpected performance.

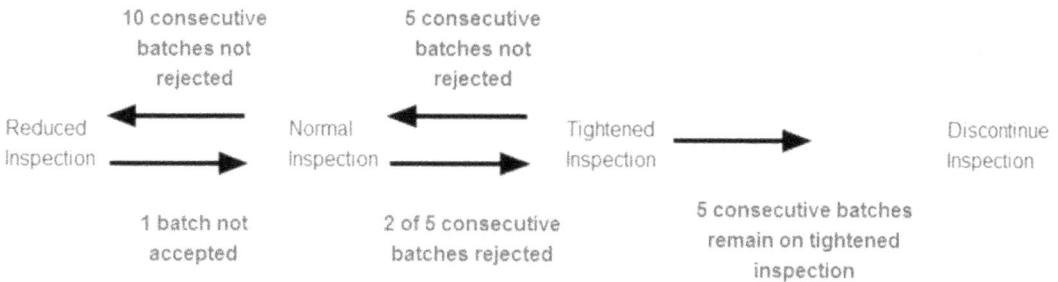

Figure 2.10: Schematic of
switching rules for sampling
by attributes

Sampling Plan Tables

With the code letter and required AQL in hand, we search for
the table with the wanted inspection type. Continuing with our
example, we use the normal inspection table shown in Figure
2.11. Using code letter J and AQL=1% we obtain the plan $n=80$,
$Ac=2$, and $Re=3$. This means that the acceptance number is $Ac=2$
("accept the batch if 2 or less non-conforming items are found"),
and the rejection number is $Re=3$ ("reject the batch if 3 or more
non-conforming items are found"). The *rejection number* (Re) is
almost always set to $Ac+1$. The only case when the rejection
number is not equal to $Ac+1$ is in reduced sampling. In that case,

if the number of non-conforming items is between the acceptance and rejection numbers the batch is accepted, but the type of inspection for the next batch changes to normal inspection (as shown in the left-most bottom arrow in Figure 2.11).

To find a plan when an arrow is encountered, follow the arrow to the appropriate row and obtain the sample size and Ac and Re numbers from the new row. For example, in our example, for AQL=0.4% we follow the arrow down and obtain the plan n=125, Ac=1, Re=2.

Online Plans: SQCOnline.com

Obtaining sampling plans using the ISO standard is somewhat confusing and error-prone. Finding the code letter, then the appropriate inspection table, reading the plan parameters (using the arrows and reading footnotes), and understanding the resulting decision rules can be taxing to users and can easily lead to mistakes.

A more user-friendly and error-proof solution is to use the website www.sqconline.com. The user is requested to enter four process parameters: batch size, AQL, inspection level (I-III, S1-S4), and type of inspection (normal/reduced/tightened), using pull-down menus. If the user is not sure about the meaning of a certain parameter, they can click on "more info" to obtain more details (see Figure 2.12).   The output page includes three components (see Figure 2.13):

1. The sampling plan in terms of a decision rule (for both single-stage and double-stage plans)

2. A notice about the need to use switching rules

3. The OC curve for the single-stage plan and another plot (ASN curve) for the double-stage plan, accompanied by explanations.

The output screen shown in Figure 2.13 gives the plan for our tire example. It also shows the OC curve for the plan (bottom right plot).

Table II-A. Single Sampling Plans for Normal Inspection (Master Table).

Legend: ↓ = Use first sampling plan below arrow. ↑ = Use first sampling plan above arrow. Ac = Acceptance number. Re = Rejection number. Each cell shows "Ac Re".

Sample size code letter	Sample size	0.010	0.015	0.025	0.040	0.065	0.10	0.15	0.25	0.40	0.65	1.0	1.5	2.5	4.0	6.5	10	15	25	40	65	100	150	250	400	650	1000
A	2	↓	↓	↓	↓	↓	↓	↓	↓	↓	↓	↓	↓	↓	↓	↓	↓	0 1	1 2	2 3	3 4	5 6	7 8	10 11	14 15	21 22	↑
B	3	↓	↓	↓	↓	↓	↓	↓	↓	↓	↓	↓	↓	↓	↓	↓	0 1	1 2	2 3	3 4	5 6	7 8	10 11	14 15	21 22	↑	↑
C	5	↓	↓	↓	↓	↓	↓	↓	↓	↓	↓	↓	↓	↓	↓	0 1	1 2	2 3	3 4	5 6	7 8	10 11	14 15	21 22	↑	↑	↑
D	8	↓	↓	↓	↓	↓	↓	↓	↓	↓	↓	↓	↓	↓	0 1	1 2	2 3	3 4	5 6	7 8	10 11	14 15	21 22	↑	↑	↑	↑
E	13	↓	↓	↓	↓	↓	↓	↓	↓	↓	↓	↓	↓	0 1	1 2	2 3	3 4	5 6	7 8	10 11	14 15	21 22	↑	↑	↑	↑	↑
F	20	↓	↓	↓	↓	↓	↓	↓	↓	↓	↓	↓	0 1	1 2	2 3	3 4	5 6	7 8	10 11	14 15	21 22	↑	↑	↑	↑	↑	↑
G	32	↓	↓	↓	↓	↓	↓	↓	↓	↓	↓	0 1	1 2	2 3	3 4	5 6	7 8	10 11	14 15	21 22	↑	↑	↑	↑	↑	↑	↑
H	50	↓	↓	↓	↓	↓	↓	↓	↓	↓	0 1	1 2	2 3	3 4	5 6	7 8	10 11	14 15	21 22	↑	↑	↑	↑	↑	↑	↑	↑
J	80	↓	↓	↓	↓	↓	↓	↓	↓	0 1	1 2	2 3	3 4	5 6	7 8	10 11	14 15	21 22	↑	↑	↑	↑	↑	↑	↑	↑	↑
K	125	↓	↓	↓	↓	↓	↓	↓	0 1	1 2	2 3	3 4	5 6	7 8	10 11	14 15	21 22	↑	↑	↑	↑	↑	↑	↑	↑	↑	↑
L	200	↓	↓	↓	↓	↓	↓	0 1	1 2	2 3	3 4	5 6	7 8	10 11	14 15	21 22	↑	↑	↑	↑	↑	↑	↑	↑	↑	↑	↑
M	315	↓	↓	↓	↓	↓	0 1	1 2	2 3	3 4	5 6	7 8	10 11	14 15	21 22	↑	↑	↑	↑	↑	↑	↑	↑	↑	↑	↑	↑
N	500	↓	↓	↓	↓	0 1	1 2	2 3	3 4	5 6	7 8	10 11	14 15	21 22	↑	↑	↑	↑	↑	↑	↑	↑	↑	↑	↑	↑	↑
P	800	↓	↓	↓	0 1	1 2	2 3	3 4	5 6	7 8	10 11	14 15	21 22	↑	↑	↑	↑	↑	↑	↑	↑	↑	↑	↑	↑	↑	↑
Q	1250	↓	↓	0 1	1 2	2 3	3 4	5 6	7 8	10 11	14 15	21 22	↑	↑	↑	↑	↑	↑	↑	↑	↑	↑	↑	↑	↑	↑	↑
R	2000	↓	0 1	1 2	2 3	3 4	5 6	7 8	10 11	14 15	21 22	↑	↑	↑	↑	↑	↑	↑	↑	↑	↑	↑	↑	↑	↑	↑	↑

Acceptable Quality Levels (normal inspection)

↓ = Use first sampling plan below arrow. If sample size equals or exceeds lot or batch size, do 100 percent inspection.
↑ = Use first sampling plan above arrow.
Ac = Acceptance number.
Re = Rejection number.

Figure 2.11: Table for normal inspection by attributes

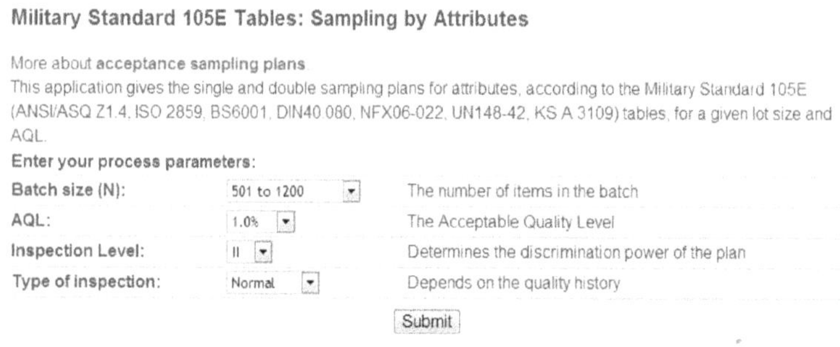

Military Standard 105E Tables: Sampling by Attributes

More about acceptance sampling plans

This application gives the single and double sampling plans for attributes, according to the Military Standard 105E (ANSI/ASQ Z1.4, ISO 2859, BS6001, DIN40.080, NFX06-022, UN148-42, KS A 3109) tables, for a given lot size and AQL.

Enter your process parameters:

Batch size (N):	501 to 1200 ▾	The number of items in the batch
AQL:	1.0% ▾	The Acceptable Quality Level
Inspection Level:	II ▾	Determines the discrimination power of the plan
Type of inspection:	Normal ▾	Depends on the quality history

Submit

Figure 2.12: Input screen on SQCOnline.com for Mil-Std-105 standard

2.5   Accept-On-Zero (AOZ) Plans

In recent years, *Accept-on-Zero* (AOZ) plans have become popular. These plans, also known as zero-defective plans, are acceptance sampling plans with acceptance threshold of $c=0$. In other words, they dictate batch acceptance only if no non-conforming items are found in the sample.

The intention behind AOZ plans is to "protect the consumer" by creating the impression that batches with non-conforming items are unacceptable, thereby putting pressure on the producer to produce high quality items. This was the intention of the U.S. Department of Defense (DOD) when it issued Mil-Std-1916 on April 1996 which contains AOZ plans for attributes, variables, and continuous sampling. Unfortunately, a different intention has led to the popularity of AOZ plans: companies who are sensitive to legal litigation by their customers (such as automotive and pharmaceutical companies) perceive AOZ plans as legal protection due to avoiding evidence of passing non-conforming items to the consumer. However, it is clear that technically AOZ plans do not offer better consumer protection compared to $c>0$ plans that satisfy the same LTPD or AQL criterion. The main criticism of AOZ plans is the misconception that they create that a sample with zero non-conforming items implies a perfect batch (which has no non-conforming items). Obviously, this is not the case. While AOZ plans are popular, they are quite controversial. From a technical point of view, they present an advantage in terms of requiring smaller samples compared to corresponding

Military Standard 105E (ANSI/ASQC Z1.4, ISO 2859) Tables

For a lot of **501 to 1200** items, **AQL=1%**, and inspection level II, the **normal** inspection plans are:

The *Single* **sampling procedure is:**

Sample **80*** items

If the number of non-conforming items is:

2 or less --> accept the lot

3 or more --> reject the lot

The *Double* **sampling procedure is:**

	Sample **50*** items
	If the number of non-conforming items is:
Step 1:	**0** --> accept the lot
	3 or more --> reject the lot
	Otherwise, continue to step 2.
	Sample **50*** additional items
	If the total number of non-conforming items is:
Step 2:	**3 or less** --> accept the lot
	4 or more --> reject the lot

Important: **This sampling plan will yield valid results only if applied with the** Military Standard 105E **switching rules.**

* Note: If the sample size(s) exceeds the lot size, apply 100% sampling of the lot.

OC Curve

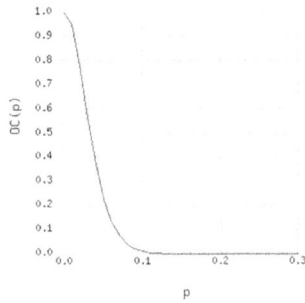

The OC curve describes the probability of accepting a lot, OC(p), as a function of the proportion non-conforming (p), for the single and double sampling plans described above.

ASN Curve

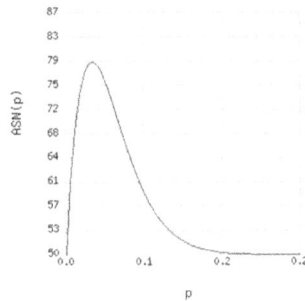

The ASN curve describes the average sample size, when employing double sampling, as a function of the proportion of non-conforming items (p).

Figure 2.13: Output screen of SQCOnline.com, giving an understandable plan and decision rule, and presenting charts with explanations

$c{>}0$ plans. However, the price of a smaller sample is the low-ered discrimination rate between good and bad lots, which in this case translates into a higher producer risk. Graphically, OC curves associated with $c{=}0$ plans have a maximum at $p{=}0$ from which they drop faster than $c{>}0$ plans. Hence, the $c{=}0$ batch acceptance probability drops fastest at the best quality levels.

In practice, there exist several AOZ standards and tables. SQCOnline.com provides calculators for obtaining AOZ plans using the popular Mil-Std-1916 standard as well as AQL-based plans using Squeglia's tables. For detailed information on the different AOZ tables and schemes see Chapter 17 in Schilling & Neubauer's book[2].

[2] *Acceptance Sampling in Quality Control*, 2nd edition, CRC Press, 2009

2.6   Single-Stage Plans for Multiple Defects-per-Unit

Thus far, we discussed the case of fail/pass data, where an item is either conforming or non-conforming. Pass/fail data are very common, because inspection often involves comparing an item to specifications or requirements and determining its conformance. Pass/fail labels are sometimes based on measuring multiple dimensions of an item. For example, a baked good might be inspected for taste, freshness, and aesthetics.

In some applications, rather than labeling an item as conform-ing or non-conforming, interest focuses on the *number of defects per unit*. For example, a medical transcription service might in-spect the number of errors per dictated recording.

Single-stage sampling plans for the number of defects consist of the sample size (n) of items to be drawn, and an acceptance number (c). Once the sample is drawn, we count the number of defects and compute defects-per-hundred-units (DPHU) as follows:

$$DPHU = \text{Defects-Per-Hundred-Units} = 100 \times \text{number of defects}/n.$$

DPHU is then compared to the acceptance number c, and the batch is accepted or rejected accordingly.

Unlike in the pass/fail case, the number of defects-per-hundred-units in the sample might exceed the sample size. Hence, AQL levels can exceed 100%. For example, if two or less defects per item are considered acceptable, then 200 defects per hundred units would be acceptable, leading to AQL=200%. The ANSI/ASQC Z1.4 tables that we used for pass/fail items provide sampling plans for multiple defects-per-units as well. The plans for AQL values up to 1,000 (meaning 1,000 defects per 100 units) are suitable for defect-per-unit inspection.

To illustrate how a sampling plan for multiple defects-per-unit is obtained from the ANSI/ASQC Z1.4 standard, consider the medical transcription example. Suppose that the service provides a hospital with 500 transcriptions daily, and one error per report is considered acceptable. Hence, AQL=100% and i=500. Using normal inspection and level II we obtain the plan n=13, c=21. Let us assume that on a certain day we inspected a random sample of 13 reports and discovered 9 errors. We compute defects-per-hundred-units as DPHU $= 100 \times 9/13 = 69.23$. Because this number exceeds the acceptance number, we reject the entire batch of 500 reports.

2.7   Sampling Plans for Specified Producer and Consumer Risks

Using the ANSI/ASQC Z1.4 standard guarantees a producer's risk (α) in the range 1%-10% for the required AQL. The reason is that the military standard was aimed at helping the US military evaluate the quality of parts from many suppliers. For a given plan, one can compute the consumer's risk by computing $\beta = OC(LTPD)$, but ANSI/ASQC Z1.4 does not give an easy way to find a plan that guarantees a particular β.

In some cases the producer and consumer require a plan that satisfies two criteria: it assures that AQL-quality batches will be accepted with high probability (low α) and LTDP-quality batches will be rejected with high probability (low β). The plan is intended to be used for a long-term relationship, thereby protecting both the producer and the consumer's interests. The two equations that one needs to solve to obtain a sampling plan of this type are:

$$1 - \alpha = OC(AQL) = P(X \leq c | p = AQL)$$

$$\beta = OC(LTPD) = P(X \leq c | p = LTPD).$$

In other words, we are looking for a sampling plan (n, c) that has an OC curve that includes the points (AQL,1-α) and (LTPD,β). Solving the two equations requires computing cumulative Binomial probabilities such as with Excel 2010's *BINOM.DIST(c,n,AQL,1)* and *BINOM.DIST(c,n,LTPD,1)*. It is possible to create an optimization program that finds n and c that solve these two equations (at least approximately, because n and c must be integers). A simpler way is to use an approximation where the Binomial probabilities are approximated by normal probabilities. The approximation leads to the solution:

$$n = \left(\frac{z_{1-\alpha}\sqrt{AQL(1-AQL)} + z_{1-\beta}\sqrt{LTPD(1-LTPD)}}{AQL - LTPD} \right)^2$$

(2.5)

$$c = nAQL - 0.5 + z_{1-\alpha}\sqrt{nAQL(1-AQL)}$$

(2.6)

The notation $z_{1-\alpha}$ denotes the $1 - \alpha$ percentile from the standard normal distribution (computed in Excel 2010 as *NORM.S.INV(1-α)*. Note that the resulting n and c must be rounded to the nearest integers. It is then advisable to compute $OC(AQL)$ and $OC(LTPD)$ for the final plan (n, c) to obtain the actual values of α and β after the rounding. This normal approximation is sufficiently accurate when $(n)(AQL) > 5$.

   For example, let us assume that we require a producer's risk of α=0.05 for quality AQL=0.015, and consumer's risk β=0.10 for quality LTPD=0.08. The two equations (2.5)-(2.6) give n=70.98, c=2.24. After rounding these numbers to obtain the plan n=71 and c=3 we compute the resulting risks and obtain α=0.02 and β=0.17, which are slightly different from the required risks, in favor of the producer. The reason for the deviation is because the criterion $(n)(AQL) > 5$ is not satisfied with this plan.

2.8   Problems

1. *Bee Healthy* is a coop of honey collectors in India. The coop regularly inspects batches of their honey jars to assure that quality is up to the standards of an organic certifying agency. A honey jar can either be conforming or non-conforming.

Each batch consists of 300 jars. The coop employs a sampling plan with $n=100$ and $c=2$.

(a) What is the Operating Characteristics (OC) for a single batch, if the quality level is $p=0.02$?

(b) What is the Operating Characteristics (OC) for a large number of batches, if the quality level is $p=0.02$?

(c) What is the producer's risk if AQL=0.01?

(d) What is the consumer's risk if LTPD=0.04?

(e) Plot the OC curve for the sampling plan.

(f) What will happen to the producer and consumer's risks if the acceptance number is increased to $c=4$?

(g) What will happen to the producer and consumer's risks if the sample size is increased to $n=200$?

(h) Find the ANSI/ASQC Z1.4 single-stage sampling plan that guarantees AQL=0.01 (use normal inspection and level II inspection), and describe it in words.

2. A hospital purchases surgical gloves from two suppliers. Supplier A has quality level $p=0.01$ and supplier B has $p=0.03$.

(a) Suppose that the hospital applies the same sampling plan of $n=200$, $c=4$ to both suppliers. What is the long term probability of accepting batches of gloves from each of the suppliers?

(b) If the hospital applies a new sampling plan with $n=50$, $c=1$ to supplier A, what is the long term probability of accepting batches of gloves from this supplier?

(c) The two sampling plans ($n = 200, c = 4$) and ($n = 50, c = 1$) have the same ratio between sample size and acceptance number. Why is there a difference between the acceptance probabilities for supplier A that you computed in (a) and (b)?

3. *Save Trees* is a vendor of recycled paper. It receives batches of 2500 pages of recycled paper from a recycling facility every week. Each sheet of paper can either be conforming or

non-conforming. To assure that the quality of the recycled paper is sufficient, *Save Trees* would like to start deploying acceptance sampling, using an ISO sampling plan. The quality requirement is AQL=0.25%.

(a) Find the ANSI/ASQC Z1.4 single-stage sampling plan, and describe it in words.

(b) Using the paper tables, what is the code letter for the above plan?

(c) If for five consecutive weeks the batches of recycled paper are accepted, what plan should *Save Trees* apply the following week?

(d) Create OC curves for the sampling plans in (a) and (c) and overlay them on the same plot. Use increments of 0.005 for p. How do the two curves differ?

(e) Find the points on the normal plan's OC curve that correspond to AQL=0.5% and LTPD=1%. What are the resulting acceptance probabilities? What are the producer and consumer risks?

(f) Use the normal approximation formulas to compute n and c that give risks equal to those found in (e) for AQL=0.5% and LTPD=1%. Report the unrounded and rounded results.

3 Double-Stage Inspection Plans for Attributes

3.1 Double-Stage Sampling Procedure

Double-stage sampling plans involve drawing a second sample from the batch when the results from the first sample are inconclusive. More formally, a double sampling plan for pass/fail data prescribes the following procedure:

- Draw a random sample of size n_1 from the batch.

- Compare the number of non-conforming items in the sample X_1 to the acceptance number c_1 and rejection number r_1.

- If $X_1 \leq c_1$ then accept the batch. If $X_1 \geq r_1$ then reject the batch. Otherwise, draw another random sample of size n_2 from the batch.

- Compare the *total number* of non-conforming items in both samples $X_1 + X_2$ to the acceptance number c_2.

- If $X_1 + X_2 \leq c_2$ then accept the batch. Otherwise, reject it.

Hence, a double-stage plan consists of five parameters: n_1, n_2, c_1, r_1, and c_2. It is common to simplify the plan by using equal sample sizes in the two stages ($n_1 = n_2$).

3.2 Why Use Double-Stage Plans?

The main advantage of double-stage plans compared to single-stage plans with similar performance is reduced amount of inspection. The single-stage sample size n is larger than each of the double-stage sample sizes n_1 and n_2. Hence, when the first sample provides sufficient evidence for acceptance or rejection of the batch, the amount of inspection is smaller.

When does the first sample provide sufficient evidence? There are two cases: when the batch quality is very low and when it is very high. In both cases the first sample is likely to reach a final decision regarding batch acceptance or rejection.

Note that the reduction in inspection efforts in double-stage sampling comes at a cost: administering a double-stage sample is more involved and in some cases can be more costly. Hence, the decision between a single-stage and double-stage sampling should take into account not only the cost of inspection but also the costs of administering each of the two plan types.

3.3   Probability of Acceptance and OC Curves

As in single-stage plans, we are interested in computing the probability of accepting the batch when using a certain plan, given different quality levels. We denoted this probability by $OC(p)$.

A batch can be accepted either in stage 1 (If $X_1 \leq c_1$) or in stage 2 (if $c_1 < X_1 < r_1$ and $X_1 + X_2 \leq c_2$). Hence, the probability of acceptance can be written as

$$
\begin{aligned}
OC(p) &= P(\text{accept in stage 1}) + P(\text{no decision in stage 1, then accept in stage 2}) \\
&= P(X_1 \leq c_1) + P(c_1 < X_1 < r_1 \text{ AND } X_1 + X_2 \leq c_2) \qquad (3.1)
\end{aligned}
$$

Example: Automobile Tires

Recall the example of the Mexican automobile tire supplier who ships batches of size $N=1,000$ to a Canadian retailer (described in Section 2.1). Consider the double-stage sampling plan with parameters $n_1 = n_2 = 50$, $c_1 = 0$, $r_1 = 3$, and $c_2 = 3$. This plan means that we start by drawing a sample of 50 tires. If all tires are conforming ($X_1 = 0$), then we accept the batch and stop inspection. If there are 3 or more non-conforming tires ($X_1 \geq 3$), we reject the batch and cease inspection. If the number of non-conforming tires is 1 or 2, we draw another sample of 50 tires. Then, we compare the total number of non-conforming tires from the 100 tires that we inspected to 3. If the total exceeds 3 non-conforming items ($X_1 + X_2 > 3$) we reject the batch. Otherwise, we accept it. Note that if the number of non-conforming items in the first stage is 1 or 2, then to accept the batch in the second stage we must find only 2 or 1 additional non-conforming tires, respectively (otherwise, we exceed the acceptance number $c_2 = 3$).

The probability of accepting the batch using this plan can be written using equation (3.1):

$$\begin{aligned} OC(p) &= P(X_1 = 0) + P(0 < X_1 < 3 \text{ AND } X_1 + X_2 \le 3) \\ &= P(X_1 = 0) + P(X_1 = 1 \text{ AND } X_2 \le 2) + P(X_1 = 2 \text{ AND } X_2 \le 1) \end{aligned}$$

To simplify this formula, we must recall that the two samples are drawn independently from the batch. Hence, the number of non-conforming tires in the first sample (event A) is independent of the number of non-conforming tires in the second sample (event B). This independence means that we can write the probabilities P(event A AND event B) as a multiplication of the form P(event A) × P(event B):

$$OC(p) = P(X_1 = 0) + P(X_1 = 1)P(X_2 \le 2) + P(X_1 = 2)P(X_2 \le 1)$$

Each of these probabilities can now be easily computed using Excel or an online calculator, using the Binomial distribution. For example, for $p=0.01$ we obtain

$$OC(0.01) = 0.605 + 0.306 \times 0.986 + 0.076 \times 0.911 = 0.975.$$

Using the same logic, we obtained the OC curve shown in Figure 3.1. We overlaid this curve with the OC curve for the single-stage plan obtained in Section 2.4 ($n=80$, $c=2$). We can see that the two plans have very similar performance in terms of batch acceptance.

Note: We used Binomial calculations assuming a long-term relationship between the supplier and retailer (infinite batch size). If we were only interested in per-batch performance, we would use Hypergeometric calculations.

OC Curves

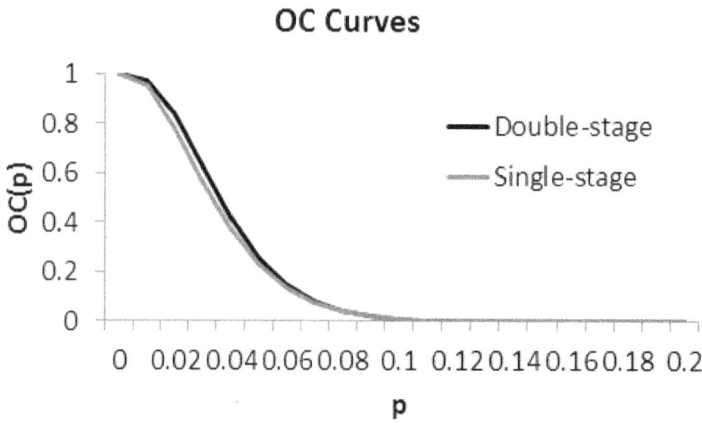

Figure 3.1: OC curves for the double-stage plan ($n_1 = n_2 = 50, c_1 = 0, r_1 = 3, c_2 = 3$) and single stage plan ($n = 80, c = 3$)

3.4   Expected Amount of Inspection

The amount of inspection performed in double-stage sampling depends on whether we reach a decision after the first sample. In general, the amount of inspection will either be n_1 or $n_1 + n_2$. For a certain quality level p, we can compute the expected amount of inspection, called *Average Sample Number* (ASN):

$$ASN(p) = n_1 + n_2 P(\text{no decision in stage 1}) = n_1 + n_2 P(c_1 < X_1 < r_1).$$
$$(3.2)$$

Note that the ASN gives the expected number of inspected units per batch over many batches, rather than for a single batch. For instance, for the double-stage plan described in the previous section ($n_1 = n_2 = 50, c_1 = 0, r_1 = 3$, and $c_2 = 3$), the average number of tires that will be inspected per batch if the quality level is 1% non-conforming tires is

$$
\begin{aligned}
ASN(0.01) &= 50 + 50P(0 < X_1 < 3) = 50 + 50[P(X_1 = 1) + P(X_1 = 2)] \\
&= 50 + 50(0.306 + 0.076) = 69.
\end{aligned}
$$

In other words, if the supplier's quality is 1%, then employing the double sampling plan will require inspecting 69 tires per batch, on average. Compared to the single-stage plan $n = 80, c = 2$ that has similar performance in terms of batch acceptance (as shown in Figure 3.1), the double-stage plan requires on average 11 less tires to inspect.

We can also plot an ASN curve, to see how the amount of inspection is expected to change as a function of batch quality. Figure 3.2 shows this curve. We can see that beyond p=0.14 we are most likely to inspect only 50 tires. The worst case is at p=0.03, where $ASN(0.03) = 80$.

ASN Curve for Double-Stage Plan

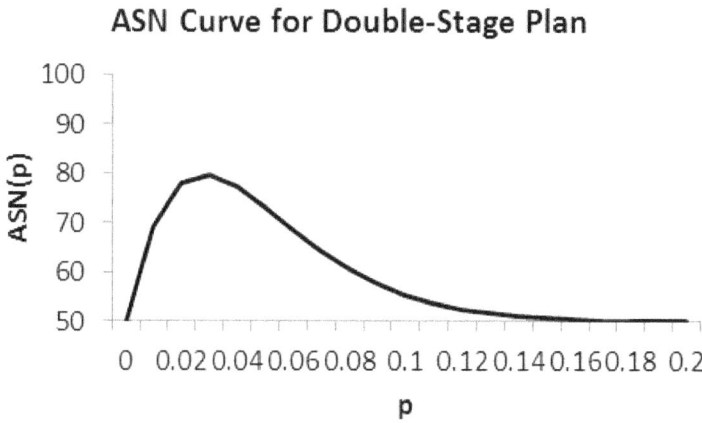

Figure 3.2: Expected amount of inspection per batch (N=1,000) using double-stage sampling

3.5   ANSI/ASQC Z1.4 Tables (Mil-Std105E)

As in the case of single-stage plans, the most widely used double-stage inspection plans for attributes are taken from the ANSI/ASQC Z1.4 standard.

Table Usage

The procedure for using the tables is similar to the single-stage procedure in terms of determining the code letter (based on batch size and inspection level). One then chooses a *Double Sampling Plan* table with the required inspection type (normal/tightened/reduced). The table gives the sample sizes and acceptance and rejection numbers. For example, for the automobile tire inspection example, let us search for a double-stage plan for batches of size $N=1{,}000$ and AQL=1%. As in the single-stage plan, the code letter is J (see Figure 2.9). We then look up the table Double Sampling Plans for Normal Inspection (see Figure 3.3) and find the plan $n_1 = n_2 = 50$ (in the third column from the left) and the acceptance and rejection numbers are given in the form $\begin{matrix} 0 & 3 \\ 3 & 4 \end{matrix}$. The top row corresponds to the first stage acceptance and rejection numbers, respectively. The bottom row corresponds to the second stage acceptance and rejection numbers, respectively. Hence we get $c_1=0$, $r_1=3$, and $c_2=3$.

As in single-stage sampling, the arrows in the table direct us to the sampling plan to use. When they lead to a dot, there is no double-stage plan available and instead we should use the single-stage plan.

Note that as in single-stage plans, the ISO standard guarantees a producer's risk (α) between 1%-10%. In addition, the double-stage plan for a given code letter and AQL level is the one that best matches the corresponding single-stage plan in terms of the OC curve.

Online Plans

Double-stage plans are easily obtained using SQCOnline.com. In fact, there is no separate procedure for obtaining the single-stage and double-stage plans. Once the initial details are entered into the input screen (Figure 2.12) the output screen contains both the single-stage plan and the double-stage plan (if one exists). This can be seen in Figure 2.13, which displays the double-sampling plan that we just looked up for the tire example.

The output screen also produces an ASN curve for the double-sampling scheme. The reason that there is only one OC curve is that it is common to both the single and double stage plans.

Table III - A. Double Sampling Plans for Normal Inspection (Master Table).

Sample size code letter	Sample	Sample size	Cumulative sample size	Acceptable Quality Levels (normal inspection)																									
				0.010	0.015	0.025	0.040	0.065	0.10	0.15	0.25	0.40	0.65	1.0	1.5	2.5	4.0	6.5	10	15	25	40	65	100	150	250	400	650	1000
				Ac Re	Ac Re	Ac Re	Ac Re	Ac Re	Ac Re	Ac Re	Ac Re	Ac Re	Ac Re	Ac Re	Ac Re	Ac Re	Ac Re	Ac Re	Ac Re	Ac Re	Ac Re	Ac Re	Ac Re	Ac Re	Ac Re	Ac Re	Ac Re	Ac Re	Ac Re
A				↓	↓	↓	↓	↓	↓	↓	↓	↓	↓	↓	↓	↓	↓	•	↓	↓	•	•	•	•	•	•	•	•	•
B	First	2	2	↓	↓	↓	↓	↓	↓	↓	↓	↓	↓	↓	↓	↓	•	↑	↓	0 2	0 3	1 4	2 5	3 7	5 9	7 11	11 16	17 22	25 31
	Second	2	4																	1 2	3 4	4 5	6 7	8 9	12 13	18 19	26 27	37 38	56 57
C	First	3	3	↓	↓	↓	↓	↓	↓	↓	↓	↓	↓	↓	↓	•	↑	↓	0 2	0 3	1 4	2 5	3 7	5 9	7 11	11 16	17 22	25 31	↑
	Second	3	6																1 2	3 4	4 5	6 7	8 9	12 13	18 19	26 27	37 38	56 57	
D	First	5	5	↓	↓	↓	↓	↓	↓	↓	↓	↓	↓	↓	•	↑	↓	0 2	0 3	1 4	2 5	3 7	5 9	7 11	11 16	17 22	25 31	↑	↑
	Second	5	10															1 2	3 4	4 5	6 7	8 9	12 13	18 19	26 27	37 38	56 57		
E	First	8	8	↓	↓	↓	↓	↓	↓	↓	↓	↓	↓	•	↑	↓	0 2	0 3	1 4	2 5	3 7	5 9	7 11	11 16	17 22	25 31	↑	↑	↑
	Second	8	16														1 2	3 4	4 5	6 7	8 9	12 13	18 19	26 27	37 38	56 57			
F	First	13	13	↓	↓	↓	↓	↓	↓	↓	↓	↓	•	↑	↓	0 2	0 3	1 4	2 5	3 7	5 9	7 11	11 16	↑	↑	↑	↑	↑	↑
	Second	13	26													1 2	3 4	4 5	6 7	8 9	12 13	18 19	26 27						
G	First	20	20	↓	↓	↓	↓	↓	↓	↓	↓	•	↑	↓	0 2	0 3	1 4	2 5	3 7	5 9	7 11	11 16	↑	↑	↑	↑	↑	↑	↑
	Second	20	40												1 2	3 4	4 5	6 7	8 9	12 13	18 19	26 27							
H	First	32	32	↓	↓	↓	↓	↓	↓	↓	•	↑	↓	0 2	0 3	1 4	2 5	3 7	5 9	7 11	11 16	↑	↑	↑	↑	↑	↑	↑	↑
	Second	32	64											1 2	3 4	4 5	6 7	8 9	12 13	18 19	26 27								
J	First	50	50	↓	↓	↓	↓	↓	↓	•	↑	↓	0 2	0 3	1 4	2 5	3 7	5 9	7 11	11 16	↑	↑	↑	↑	↑	↑	↑	↑	↑
	Second	50	100										1 2	3 4	4 5	6 7	8 9	12 13	18 19	26 27									
K	First	80	80	↓	↓	↓	↓	↓	•	↑	↓	0 2	0 3	1 4	2 5	3 7	5 9	7 11	11 16	↑	↑	↑	↑	↑	↑	↑	↑	↑	↑
	Second	80	160									1 2	3 4	4 5	6 7	8 9	12 13	18 19	26 27										
L	First	125	125	↓	↓	↓	↓	•	↑	↓	0 2	0 3	1 4	2 5	3 7	5 9	7 11	11 16	↑	↑	↑	↑	↑	↑	↑	↑	↑	↑	↑
	Second	125	250								1 2	3 4	4 5	6 7	8 9	12 13	18 19	26 27											
M	First	200	200	↓	↓	↓	•	↑	↓	0 2	0 3	1 4	2 5	3 7	5 9	7 11	11 16	↑	↑	↑	↑	↑	↑	↑	↑	↑	↑	↑	↑
	Second	200	400							1 2	3 4	4 5	6 7	8 9	12 13	18 19	26 27												
N	First	315	315	↓	↓	•	↑	↓	0 2	0 3	1 4	2 5	3 7	5 9	7 11	11 16	↑	↑	↑	↑	↑	↑	↑	↑	↑	↑	↑	↑	↑
	Second	315	630						1 2	3 4	4 5	6 7	8 9	12 13	18 19	26 27													
P	First	500	500	↓	•	↑	↓	0 2	0 3	1 4	2 5	3 7	5 9	7 11	11 16	↑	↑	↑	↑	↑	↑	↑	↑	↑	↑	↑	↑	↑	↑
	Second	500	1000					1 2	3 4	4 5	6 7	8 9	12 13	18 19	26 27														
Q	First	800	800	•	↑	↓	0 2	0 3	1 4	2 5	3 7	5 9	7 11	11 16	↑	↑	↑	↑	↑	↑	↑	↑	↑	↑	↑	↑	↑	↑	↑
	Second	800	1600				1 2	3 4	4 5	6 7	8 9	12 13	18 19	26 27															
R	First	1250	1250	↑	↑	0 2	0 3	1 4	2 5	3 7	5 9	7 11	11 16	↑	↑	↑	↑	↑	↑	↑	↑	↑	↑	↑	↑	↑	↑	↑	↑
	Second	1250	2500			1 2	3 4	4 5	6 7	8 9	12 13	18 19	26 27																

↓ = Use first sampling plan below arrow. If sample size equals or exceeds lot or batch size, do 100 percent inspection.
↑ = Use first sampling plan above arrow.
• = Use corresponding single sampling plan (or alternatively, use double sampling plan below, where available).

Ac = Acceptance number.
Re = Rejection number.

Figure 3.3: ANSI/ASQC Z1.4 table for double-stage sampling plans, normal inspection

3.6   Problems

1. *Bee Healthy* is a coop of honey collectors in India. They regularly inspect batches of their honey jars to assure that their quality is up to the standards of an organic certifying agency. A honey jar can either be conforming or non-conforming. Each batch consists of 300 jars. The costs associated with inspection include Rs.40 for drawing a sample, and Rs.1 for inspecting a single jar.

   (a) Find an ANSI/ASQC Z1.4 single-stage plan that yields AQL=4% (use normal inspection, level II). Compute the inspection cost of this plan.

   (b) Find the corresponding double-stage ANSI/ASQC Z1.4 double-stage (AQL=4%, normal inspection, level II). Compute the expected inspection cost of this plan if the incoming quality is 0.05.

   (c) When is the single-stage plan more costly to implement?

   (d) What other considerations should influence *Bee Healthy*'s choice between single-stage and double-stage inspection?

2. *Save Trees* is a vendor of recycled paper. It receives batches of 2500 pages of recycled paper from a recycling facility every week. Each sheet of paper can either be conforming or non-conforming. To assure that the quality of the recycled paper is sufficient, *Save Trees* would like to start deploying acceptance sampling, using an ISO sampling plan. The quality requirement is AQL=0.25%.

   (a) Find the ANSI/ASQC Z1.4 single-stage sampling plan, and describe it in words.

   (b) Find ANSI/ASQC Z1.4 double-stage sampling plan, and describe it in words.

   (c) Create OC curves for the two plans from (a) and (b) and overlay them in the same plot.

   (d) Create the ASN curve for the double-stage plan. What is the highest expected number of pages to be inspected per batch? For which quality level (p) is this ASN obtained?

(e) When would it be advantageous for *Save Trees* to use a double-stage plan? When would a single-stage plan be preferable?

4 Rectifying Sampling Plans

In the sampling plans described in previous sections, once non-conforming items are found in the sample they are simply discarded. This is often because the inspection itself is destructive. Hence such sampling plans are called *non-rectifying plans*.

In some instances, it is possible to easily fix non-conforming items that are detected. In other cases it is too wasteful to discard non-conforming items and rectifying is used. In such cases, Dodge and Romig suggested schemes where non-conforming items are rectified or replaced.

4.1 Single-Stage Rectifying Plans

The following is called the Dodge-Romig single-stage rectifying scheme:

- From a batch of size N, draw a random sample of size n.

- Inspect the sample, and count the number of non-conforming items (X).

- If $X \leq c$, then accept the batch and rectify or replace the X non-conforming items in the sample.

- If $X > c$, inspect the entire batch and rectify or replace all the non-conforming items found.

The result of deploying this inspection scheme is that the outgoing batch has higher quality than the incoming quality p (before inspection). In particular, if the entire batch has been inspected, then the outgoing quality is perfect ($p=0$). If the batch was accepted, then it has exactly X less non-conforming items. The

Average Outgoing Quality (AOQ) of a batch that originally had a proportion p of non-conforming items, and which undergoes the rectifying inspection scheme is given by

$$
\begin{aligned}
AOQ(p) &= 0 \times P(\text{Batch Rejected}) + \frac{N-n}{N} p \times P(\text{Batch Accepted}) \\
&= \frac{N-n}{N} p \times OC(p) \approx p \times OC(p) \qquad (4.1)
\end{aligned}
$$

It is useful to plot the AOQ as a function of p, to see what the expected outgoing quality will be under different scenarios of ingoing quality.

The second measure of interest is the amount of inspection that this rectifying scheme will require. When the batch is accepted, we only inspect n items. When the batch is rejected we inspect the entire batch. Looking at it differently, in any case we inspect n items. We then inspect the remaining $N - n$ items if the batch is rejected. The *Average Total Inspection* (ATI) measure is therefore given by:

$$
ATI(p) = n + (N-n)(1 - OC(p)) \qquad (4.2)
$$

Choosing n and c

Dodge and Romig created two sets of sampling plans for protecting the consumer. The first set is aimed at achieving a required per-batch LTPD and assures a consumer risk of $\beta=0.10$. The second set is aimed at achieving a worst-case per-batch AOQ, called AOQ limit (AOQL).

In both cases, there are multiple sampling plans (pairs of n, c) that achieve the LTPD or AOQL criterion, and therefore the tables give the plan with the lowest expected amount of inspection (minimum ATI).

To use the Dodge-Romig tables, the user must also specify the *process average*, which is an estimate of the production quality. Let us illustrate the use of the two types of plans using our previous tire example. Note that the term LQL used in the table is equivalent to LTPD (the per-batch quality limit).

AOQL-Based Plans

Using the Dodge-Romig tables for obtaining an AOQL scheme sampling plan involves the following steps:

- Estimate the process quality level.

- Choose the worst-case acceptable average outgoing quality level (AOQL).

- Given the batch size N and process quality, find the pair (n, c).

Let us assume that the process quality is approximately 1%, and that the required AOQL is 2% (recall that N=1,000). Using the table in Figure 4.1, we obtain the plan n=65, c=2 with LQL=8.1%. This plan means that a sample of 65 items should be inspected. If 2 or less non-conforming items are found, the batch is accepted and the non-conforming items are rectified or replaced. Otherwise, the entire batch is inspected and all the non-conforming items are rectified or inspected.

This plan is guaranteed to achieve a maximum Average Outgoing Quality of 2% across many batches, with LTPD=8.1% for a particular batch. The average total inspection (ATI) associated with this plan can be computed using equation (4.2). Figure 4.2 shows the ATI curve for different levels of incoming quality (p).

LTPD-Based Plans

Obtaining an LTPD-based rectifying sampling plan using the Dodge-Romig tables is similar to the AOQL scheme, except for specifying the required per-batch LTPD. The steps are:

- Estimate the process quality level.

- Choose a worst-case acceptable per-batch LTPD value.

- Given the batch size N and process quality, find the pair (n, c).

For example, if we specify LTPD=2% for our tire example, then using the table in Figure 4.3 for a batch of size 1,000 and a process average of 1%, we obtain the plan n=305, c=3. The

PROCESS AVERAGE (%)

LOT SIZE	0.00 - 0.04			0.05 - 0.40			0.41 - 0.80			0.81 - 1.20			1.21 - 1.60		
	n	c	LQL (%)	n	c	LQL (%)	n	c	LQL (%)	n	c	LQL (%)	n	c	LQL (%)
1 - 15	All	0	-	All	0	-	All	0	-	All	0	-	All	0	-
16 - 50	14	0	13.6	14	0	13.6	14	0	13.6	14	0	13.6	14	0	13.6
51 - 100	16	0	12.4	16	0	12.4	16	0	12.4	16	0	12.4	16	0	12.4
101 - 200	17	0	12.2	17	0	12.2	17	0	12.2	17	0	12.2	35	1	10.5
201 - 300	17	0	12.3	17	0	12.3	17	0	12.3	37	1	10.2	37	1	10.2
301 - 400	18	0	11.8	18	0	11.8	38	1	10.0	38	1	10.0	38	1	10.0
401 - 500	18	0	11.9	18	0	11.9	39	1	9.8	39	1	9.8	60	2	8.6
501 - 600	18	0	11.9	18	0	11.9	39	1	9.8	39	1	9.8	60	2	8.6
601 - 800	18	0	11.9	40	1	9.6	40	1	9.6	65	2	8.0	65	2	8.0
801 - 1000	18	0	12.0	40	1	9.6	40	1	9.6	65	2	8.1	65	2	8.1
1001 - 2000	18	0	12.0	41	1	9.4	65	2	8.2	65	2	8.2	95	3	7.0
2001 - 3000	18	0	12.0	41	1	9.4	65	2	8.2	95	3	7.0	120	4	6.5
3001 - 4000	18	0	12.0	42	1	9.3	65	2	8.2	95	3	7.0	155	5	6.0
4001 - 5000	18	0	12.0	42	1	9.3	70	2	7.5	125	4	6.4	155	5	6.0
5001 - 7000	18	0	12.0	42	1	9.3	95	3	7.0	125	4	6.4	185	6	5.6
7001 - 10000	42	1	9.3	70	2	7.5	95	3	7.0	155	5	6.0	220	7	5.4
10001 - 20000	42	1	9.3	70	2	7.6	95	3	7.0	190	6	5.6	290	9	4.9
20001 - 50000	42	1	9.3	70	2	7.6	125	4	6.4	220	7	5.4	395	12	4.5
50001 - 100000	42	1	9.3	95	3	7.0	160	5	5.9	290	9	4.9	505	15	4.2

Figure 4.1: Dodge-Romig Single Sampling Lot Inspection Table. Based on Average Outgoing Quality Limit (AOQL)=2%

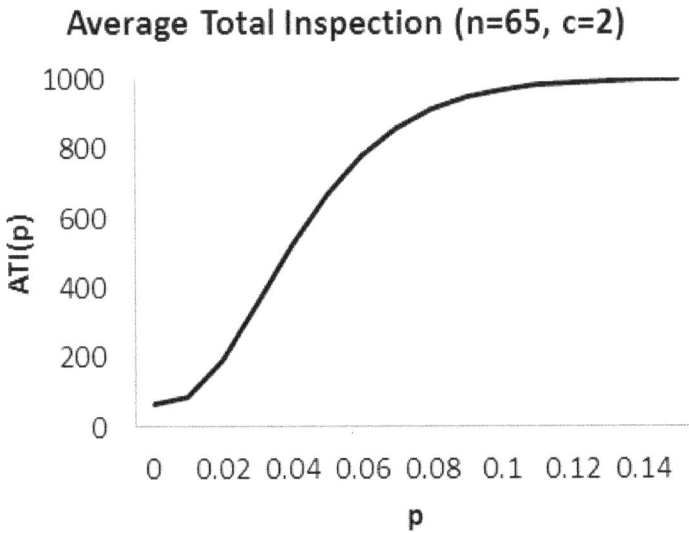

Average Total Inspection (n=65, c=2)

Figure 4.2: Average Total Inspection (ATI) curve for rectifying sampling plan $n=65, c=2$

AOQL=0.44% of this plan means that in the long run, the average outgoing quality across many batches will be no worse than 0.44%.

Online Plans: SQCOnline.com

SQCOnline.com offers a user-friendly and error-proof tool for obtaining Dodge-Romig rectifying plans, by either AOQL or LTPD. The user chooses the required AOQL or LTPD level, the lot size, and the estimated process average. The screenshot in Figure 4.4 shows an input screen with the parameters from our previous example: AOQL=2%, process average = 1%, and lot size 1,000. The output screen with the corresponding plan is shown in Figure 4.5, clearly describing the plan and how to implement it.

4.2   Double-Stage Rectifying Plans

The Dodge-Romig tables include sampling plans for double-stage schemes in addition to single-stage plans. As in the simple-

Table 2. Dodge and Romig Single Sampling Lot Inspection Table.

Based on Stated Value of Lot Tolerance Per Cent Defective (LTPD) = 2.0 % and Consumer's Risk = 0.10.

LOT SIZE	Process Average (%)																	
	0 - 0.02			0.03 - 0.20			0.21 - 0.40			0.41 - 0.60			0.61 - 0.80			0.81 - 1.00		
	n	c	AOQL (%)	n	c	AOQL (%)	n	c	AOQL (%)	n	c	AOQL (%)	n	c	AOQL (%)	n	c	AOQL (%)
1 - 75	All	0	0.00	All	0	0.00	All	0	0.00	All	0	0.00	All	0	0.00	All	0	0.00
76 - 100	70	0	0.16	70	0	0.16	70	0	0.16	70	0	0.16	70	0	0.16	70	0	0.16
101 - 200	85	0	0.25	85	0	0.25	85	0	0.25	85	0	0.25	85	0	0.25	85	0	0.25
201 - 300	95	0	0.26	95	0	0.26	95	0	0.26	95	0	0.26	95	0	0.26	95	0	0.26
301 - 400	100	0	0.28	100	0	0.28	100	0	0.28	160	1	0.32	160	1	0.32	160	1	0.32
401 - 500	105	0	0.28	105	0	0.28	105	0	0.28	165	1	0.34	165	1	0.34	165	1	0.34
501 - 600	105	0	0.29	105	0	0.29	175	1	0.34	175	1	0.34	175	1	0.34	235	2	0.36
601 - 800	110	0	0.29	110	0	0.29	180	1	0.36	240	2	0.40	240	2	0.40	300	3	0.41
801 - 1000	115	0	0.28	115	0	0.28	185	1	0.37	245	2	0.42	305	3	0.44	305	3	0.44
1001 - 2000	115	0	0.30	190	1	0.40	255	2	0.47	325	3	0.50	380	4	0.54	440	5	0.56
2001 - 3000	115	0	0.31	190	1	0.41	260	2	0.48	385	4	0.58	450	5	0.60	565	7	0.64
3001 - 4000	115	0	0.31	195	1	0.41	330	3	0.54	450	5	0.63	510	6	0.65	690	9	0.70
4001 - 5000	195	1	0.41	260	2	0.50	335	3	0.54	455	5	0.63	575	7	0.69	750	10	0.74
5001 - 7000	195	1	0.42	265	2	0.50	335	3	0.55	515	6	0.69	640	8	0.73	870	12	0.80
7001 - 10000	195	1	0.42	265	2	0.50	395	4	0.62	520	6	0.69	760	10	0.79	1050	15	0.86
10001 - 20000	200	1	0.42	265	2	0.51	460	5	0.67	650	8	0.77	885	12	0.86	1230	18	0.94
20001 - 50000	200	1	0.42	335	3	0.58	520	6	0.73	710	9	0.81	1060	15	0.93	1520	23	1.00
50001 - 100000	200	1	0.42	335	3	0.58	585	7	0.76	770	10	0.84	1180	17	0.97	1690	26	1.10

n : Size of Sample; entry of "All" indicates that each piece in lot is to be inspected.

c : Allowable Defect Number for Sample.

AOQL : Average Outgoing Quality Limit.

Figure 4.3: Dodge-Romig Single Sampling Lot Inspection Table, based on LTPD=2%

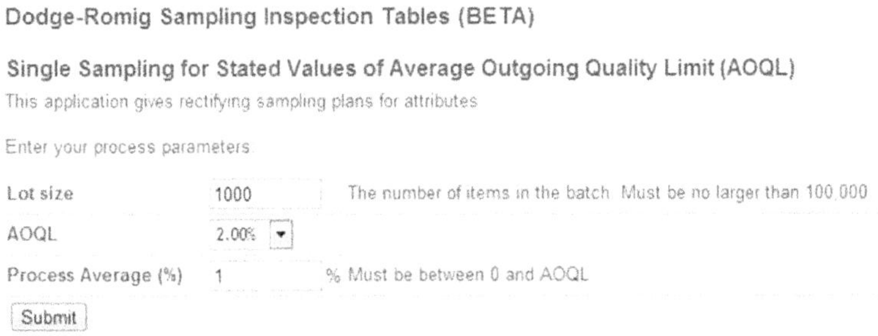

Dodge-Romig Sampling Inspection Tables (BETA)

Single Sampling for Stated Values of Average Outgoing Quality Limit (AOQL)

This application gives rectifying sampling plans for attributes

Enter your process parameters

Lot size	1000	The number of items in the batch. Must be no larger than 100,000
AOQL	2.00% ▾	
Process Average (%)	1	% Must be between 0 and AOQL

[Submit]

Dodge-Romig Sampling Inspection Tables (BETA)

Single Sampling for Stated Values of Average Outgoing Quality Limit (AOQL)

This application gives rectifying sampling plans for attributes

For a lot of **1,000** items, with AOQL **2.00%**, and process average **1%**, the inspection plan is:

Sample **65** items

If the number of non-conforming items is

- $0 \rightarrow$ accept the lot
- 1 or 2 $\rightarrow$ rectify the non-conforming item(s) and accept the lot
- 3 or more $\rightarrow$ inspect the entire lot and rectify all non-conforming items

Additional Information

p_t (quality limit for a single batch) for this process is 8.10%

Figure 4.4: Input screen on SQCOnline.com for Dodge-Romig calculator (AOQL-based plans)

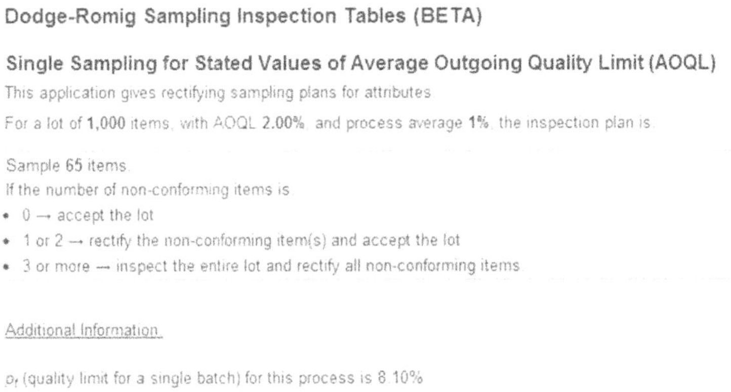

Figure 4.5: Output screen on SQCOnline.com giving the Dodge-Romig single-sample rectifying plan

stage case, there are two types of plans: AOQL-based plans and LTPD-based plans. AOQL-based plans guarantee a maximum average outgoing quality level across many batches. An LTPD-based plan guarantees a maximum outgoing quality level for a single batch.

An example of an AOQL-based double-stage table is shown in Figure 4.6. Each plan consists of the first and second stage sample sizes (n_1, n_2), a first stage acceptance number (c_1) and a second stage acceptance number (c_2). The double-stage rectifying plan follows the following procedure:

- Draw a sample of size n1 and inspect each of the items.

- If the number of non-conforming items is less than or equal to the acceptance number c_1 $(X_1 \leq c_1)$, then accept the batch and rectify or replace the X_1 items.

- Otherwise, draw another sample of size n_2, and inspect each of the items.

- If the total number of non-conforming items in the two samples is less than or equal to the acceptance number c_2 $(X_1 + X_2 \leq c_2)$ then accept the batch and rectify or replace the $X_1 + X_2$ items.

- Otherwise, inspect the entire batch and replace or rectify all non-conforming items found.

To illustrate this scheme, let us find a plan for the tire supplier that achieves AOQL=1%, where we estimate the average process quality to be 0.5%. Using the bottom table in Figure 4.6, and using the batch size of 1,000, we obtain the plan n_1=65, c_1=0, n_2=110, and c_2=3. This means that we first sample 65 tires.

- If all tires are conforming, we accept the batch and stop inspection (in this case since there is no repairing, the outgoing quality is equal to the incoming quality).

- If we find three or more non-conforming tires, we inspect the entire batch and rectify all non-conforming tires.

- If one or two non-conforming tires are found, we draw another sample of 110 tires. If the overall number of non-conforming tires (including the one or two from the first sample) exceeds 3, we inspect the entire batch and repair/replace all the non-conforming items. Otherwise, we only repair/replace the non-conforming tires that were found in the two samples.

Finally, we see that deploying this plan guarantees a per-lot LTPD of 4%.

Computing the expected number of inspected items (ATI) of rectifying double-stage plans requires considering three scenarios: inspecting only n_1 items (if $X_1 \leq c_1$), inspecting $n_1 + n_2$ items (if $X_1 > c_1$ AND $X_1 + X_2 \leq c_2$), and inspecting the entire batch (if $X_1 > c_1$ AND $X_1 + X_2 > c_2$). When there are costs associated with inspection, they can be combined with the three scenarios to produce the expected cost of inspection.

Table 10. Dodge-Romig Double Sampling Lot Inspection Table.
Based on Average Outgoing Quality Limit (AOQL) = 1.0 %.

PROCESS AVERAGE (%)

LOT SIZE	0.00 - 0.02						0.03 - 0.20						0.21 - 0.40					
	Trial 1		Trial 2			LQL	Trial 1		Trial 2			LQL	Trial 1		Trial 2			LQL
	n_1	c_1	n_2	n_1+n_2	c_2	(%)	n_1	c_1	n_2	n_1+n_2	c_2	(%)	n_1	c_1	n_2	n_1+n_2	c_2	(%)
1 - 25	All	0	-	-	-	-	All	0	-	-	-	-	All	0	-	-	-	-
26 - 50	22	0	-	-	-	7.7	22	0	-	-	-	7.7	22	0	-	-	-	7.7
51 - 100	33	0	17	50	1	6.9	33	0	17	50	1	6.9	33	0	17	50	1	6.9
101 - 200	43	0	22	65	1	5.8	43	0	22	65	1	5.8	43	0	22	65	1	5.8
201 - 300	47	0	28	75	1	5.5	47	0	28	75	1	5.5	47	0	28	75	1	5.5
301 - 400	49	0	31	80	1	5.4	49	0	31	80	1	5.4	55	0	60	115	2	4.8
401 - 500	50	0	30	80	1	5.4	50	0	30	80	1	5.4	55	0	65	120	2	4.7
501 - 600	50	0	30	80	1	5.4	50	0	30	80	1	5.4	60	0	65	125	2	4.6
601 - 800	50	0	35	85	1	5.3	60	0	70	130	2	4.5	60	0	70	130	2	4.5
801 - 1000	55	0	30	85	1	5.2	60	0	75	135	2	4.4	60	0	75	135	2	4.4
1001 - 2000	55	0	35	90	1	5.1	65	0	75	140	2	4.3	75	0	120	195	3	3.8
2001 - 3000	65	0	80	145	2	4.2	65	0	80	145	2	4.2	75	0	125	200	3	3.7
3001 - 4000	70	0	80	150	2	4.1	70	0	80	150	2	4.1	80	0	175	255	4	3.5
4001 - 5000	70	0	80	150	2	4.1	70	0	80	150	2	4.1	80	0	180	260	4	3.4
5001 - 7000	70	0	80	150	2	4.1	75	0	125	200	3	3.7	80	0	180	260	4	3.4
7001 - 10000	70	0	80	150	2	4.1	80	0	125	205	3	3.6	85	0	180	265	4	3.3
10001 - 20000	70	0	80	150	2	4.1	80	0	130	210	3	3.6	90	0	230	320	5	3.2
20001 - 50000	75	0	80	155	2	4.0	80	0	135	215	3	3.6	95	0	300	395	6	2.9
50001 - 100000	75	0	80	155	2	4.0	85	0	180	265	4	3.3	170	1	380	550	8	2.6

PROCESS AVERAGE (%)

LOT SIZE	0.41 - 0.60						0.61 - 0.80						0.81 - 1.00					
	Trial 1		Trial 2			LQL	Trial 1		Trial 2			LQL	Trial 1		Trial 2			LQL
	n_1	c_1	n_2	n_1+n_2	c_2	(%)	n_1	c_1	n_2	n_1+n_2	c_2	(%)	n_1	c_1	n_2	n_1+n_2	c_2	(%)
1 - 25	All	0	-	-	-	-	All	0	-	-	-	-	All	0	-	-	-	-
26 - 50	22	0	-	-	-	7.7	22	0	-	-	-	7.7	22	0	-	-	-	7.7
51 - 100	33	0	17	50	1	6.9	33	0	17	50	1	6.9	33	0	17	50	1	6.9
101 - 200	43	0	22	65	1	5.8	43	0	22	65	1	5.8	47	0	43	90	2	5.4
201 - 300	55	0	50	105	2	4.9	55	0	50	105	2	4.9	55	0	50	105	2	4.9
301 - 400	55	0	60	115	2	4.8	55	0	60	115	2	4.8	60	0	80	140	3	4.5
401 - 500	55	0	65	120	2	4.7	60	0	95	155	3	4.3	60	0	95	155	3	4.3
501 - 600	60	0	65	125	2	4.6	65	0	100	165	3	4.2	65	0	100	165	3	4.2
601 - 800	65	0	105	170	3	4.1	65	0	105	170	3	4.1	70	0	140	210	4	3.9
801 - 1000	65	0	110	175	3	3.8	70	0	150	220	4	3.8	125	1	180	305	6	3.5
1001 - 2000	80	0	165	245	4	3.7	135	1	200	335	6	3.2	140	1	245	385	7	3.2
2001 - 3000	80	0	170	250	4	3.6	150	1	265	415	7	3.0	215	2	355	570	10	2.8
3001 - 4000	85	0	220	305	5	3.3	160	1	330	490	8	2.8	225	2	455	680	12	2.7
4001 - 5000	145	1	225	370	6	3.1	225	2	375	600	10	2.7	240	2	595	835	14	2.5
5001 - 7000	155	1	285	440	7	2.9	235	2	440	675	11	2.6	310	3	665	975	16	2.4
7001 - 10000	165	1	355	520	8	2.7	250	2	585	835	13	2.4	385	4	785	1170	19	2.3
10001 - 20000	175	1	415	590	9	2.6	325	3	655	980	15	2.3	520	6	980	1500	24	2.2
20001 - 50000	250	2	490	740	11	2.4	340	3	910	1250	19	2.1	610	7	1410	2020	32	2.1
50001 - 100000	275	2	700	975	14	2.2	420	4	1050	1470	22	2.1	770	9	1850	2620	41	2.0

n_1 = size of sample; n_2 = size of second sample, entry of "All" indicates that each piece in lot is to be inspected
c_1 = acceptance number for sample c_2 = acceptance number for first and second samples combined
LQL = limiting quality level corresponding to a consumer's risk (β) = 0.10

Figure 4.6: Dodge-Romig table for rectifying double-stage plan, for AOQL=1%

4.3   Problems

1. A website hosting service would like to use daily rectifying sampling inspection to assure that the 3200 websites that it hosts have not been hacked. When a hacked website is detected, the hosting company quickly changes the permissions to the website and notifies the website owner. Experience has shown that on average 1% of websites are hacked daily.

    (a) Find a single-stage rectifying sampling plan that would guarantee a daily LTPD of 2%. Explain in words how the sampling plan should be deployed.

    (b) What is the long term AOQL that the hosting service can expect using this plan?

    (c) To see how many websites the hosting service should expect to inspect on each day, plot the ATI curve.

2. An airline inspects the quality of the meals to be served on its international flights using rectifying sampling inspection. Each day, a batch of 7,000 meals arrives from the catering company. A meal that is found non-conforming is immediately replaced with a conforming meal. From past experience, the airline knows that the catering company produces an average of 1% non-conforming meals.

    (a) Find a single-stage rectifying plan that achieves AOQL=2% (use the Dodge-Romig tables).

    (b) Plot the OC curve for this plan.

    (c) To see how many meals the airline should expect to inspect per batch, plot the ATI curve.

    (d) To see what outgoing quality the airline should expect, plot the AOQ curve.

    (e) If the caterer's percent of non-conforming meals is indeed 1%, what is the expected outgoing quality after applying the inspection sampling scheme? Use the AOQ curve that you plotted in (d).

3. For the airline example in problem (2), consider the following costs: drawing a sample costs $5 and inspecting a meal costs $0.25.

   (a) Find a double-stage rectifying scheme that achieves AOQL=1%, assuming that the catering company produces 0.5% non-conforming meals and explain how to carry it out in words.

   (b) What is the worst per-batch quality that this plan guarantees? What is the probability that a batch with worse quality will be accepted?

   (c) Compute the probability that a batch will be accepted after the first sample (use Binomial probabilities). Recall that the airline knows that the catering company produces an average of 1% non-conforming meals.

   (d) For a single batch of meals, compute the expected cost of inspection.

   (e) Compare the expected cost of inspecting of a single batch to the alternative of no inspection, if the penalty for serving a non-conforming meal is $10. Given the cost structure in this case, which approach is cheaper? What explains the small/large difference?

   (f) Are there additional considerations to take into account aside from costs in choosing between no inspection and the rectifying double-stage sampling?

   (g) Describe two scenarios of the airline catering relationship in which the double-stage rectifying sampling scheme can have a large advantage over no inspection.

5 Inspection Plans for Variables (Continuous Measurements)

While sampling plans for pass/fail data are very common in practice, there are also many instances in which inspection is based on a single continuous measure of interest. For example, an exclusive producer of fine chocolate inspects the weight of chocolate boxes to assure a minimum weight before shipping out to customers. A facility for water purification inspects the amount of contaminants in batches of water bottles to assure they do not exceed a dangerous value. A pharmaceutical inspects the percentage of the active ingredient in batches of a certain medication prior to distributing it to pharmacies, to assure that it is not too high or too low.

In many cases it is quicker and easier to classify an item as pass/fail than to take an exact measurement, thereby resorting to acceptance sampling for attributes. However, the advantage of exact continuous measurements over binary classification is that they contain more information. The added information translates into a need for smaller samples in acceptance sampling for variables compared to attributes. Sampling plans for continuous data are popularly termed *plans for variables*. We use the less ambiguous term *continuous data* or *continuous variables*.

Let us start by considering an example, which we will use for illustrating the different computations and issues that arise.

5.1   Example: Organic Milk Distribution

A distributor of organic cow milk would like to use acceptance
sampling to inspect the quality of incoming milk bottles. Bottles
arrive from various dairies in batches of 5,000 bottles. A milk
bottle is considered conforming if it has at least 3% fat. The
distributor requires an average quality limit (AQL) of 1%.

5.2   Specification Limits

Depending on the context, the conformance of an item is deter-
mined by one of three options:

An upper specification limit (U): a conforming measurement is not
  higher than U

A lower specification limit (L): a conforming measurement is not
  lower than L

Double specification limit (L,U): a conforming measurement
  should be between L and U

In the organic milk distribution example, the measurement of
interest is the percent of fat, and the lower specification limit is
L=3%.

5.3   Procedure

A sampling plan for continuous data follows the following general procedure:

- A sample of size n is drawn at random from the batch.

- Each of the items in the sample is inspected, yielding n measurements $(x_1, \ldots, x_n)$. The sample mean is computed by
  $\bar{x} = (x_1 + x_2 + \cdots + x_n)/n$ and standard deviation by
  $s = \sqrt{\frac{1}{n-1} \sum_{i=1}^{n}(x_i - \bar{x})^2}$.

At this point, there are two possible procedures. One is based on
measuring the distance of the sample mean from the specification limit(s). The other is based on estimating the proportion of
non-conforming items:

Procedure 1 (k-Method): Compute $Q_L = \frac{\bar{x}-L}{s}$ and/or $Q_U = \frac{U-\bar{x}}{s}$ and compare them to a lower threshold k_L and/or upper threshold k_U. If the threshold is exceeded, then the batch is rejected. Otherwise, it is accepted.

Procedure 2 (M-Method): Compute the estimated proportion of non-conforming items in the sample and compare it to an acceptance threshold M. If the acceptance threshold is exceeded, then reject the batch. Otherwise, accept it.

Procedures 1 and 2 yield identical results for single specification limits. For double specification limits only Procedure 2 (*M-Method*) can be used. Plans in ISO standards are based on choosing an AQL level and guarantee a producer's risk (α) of 1%-10%. In the following we look at AQL-based plans.

5.4   The Normality Assumption

Standard sampling plans by variables are based on a critical assumption: that the measurement distribution can be approximated by the normal distribution.

Let us denote the mean of the entire process (or batch) by μ_p and the standard deviation by σ. The term μ_p implies that the process produces a proportion p of items that exceeds specifications (L and/or U). The relationship between p and μ_p can be seen in Figure 5.1 for each the specification limit cases. For example, in the lower specification limit case (top plot), the process produces a proportion of p items below L and has a mean of μ_p.

The normality assumption means that if we plot a histogram of the measurements (such as the percentage of fat in many milk bottles) we expect to see a bell-curve shape symmetric around the process mean μ_p and which extends to approximately $\mu_p \pm 3\sigma$ (shown in Figure 5.1).

We write the assumption about the normality of the measurements using the standard notation $X \sim N(\mu_p, \sigma^2)$.

What if normality is not a reasonable assumption?

When the measurements are far from looking normally distributed, it is incorrect to use the sampling plans that are intended for normal data because they will produce incorrect
decision rules. One relatively simple solution is to try and trans-
form the measurements using functions such as logarithms and
square roots, in an effort to obtain normally-distributed measurements. For instance, often taking the logarithm of right-skewed
measurements produces measurements that are more symmetric.
Another solution is to move to pass/fail classification, where no
such assumption is made.

5.5   Known vs. Unknown Process Standard Deviation

In processes that have been operating steadily for a while, it is
common to measure the standard deviation and treat it as σ.
However, in many cases this information is unknown. In such
cases the sample standard deviation is used in place of σ in the
various formulas. There are two main differences that arise in
acceptance sampling between the case of known and unknown
σ:

1. Sampling plans for an unknown σ require larger samples than
   plans for known σ, to compensate for the uncertainty.

2. In small samples ($n < 30$), the use of normal probabilities are
   replaced with probabilities from the T distribution (see next
   section).

   In the past it was common to use the sample range (max-
min) to estimate σ because of the ease of computing the range
by hand. However, the range is an inferior estimate compared
to the sample standard deviation (s), and therefore in today's
environment where it is very fast and easy to compute s the
range is no longer used.

Figure 5.1: Normal distribution curves with mean μ_p and standard deviation σ, for the three specification limit cases

Lower specification limit

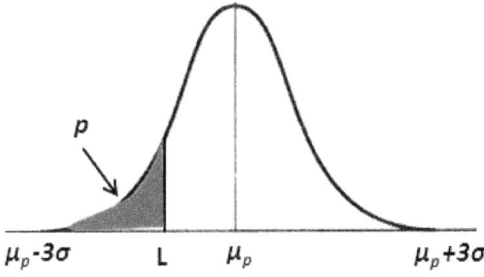

Upper specification limit

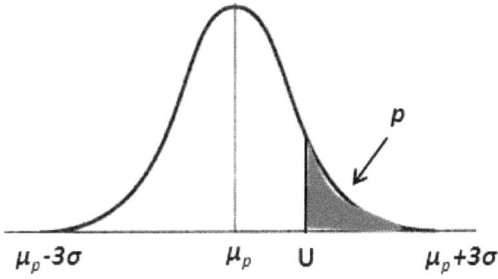

Lower & upper specification limits

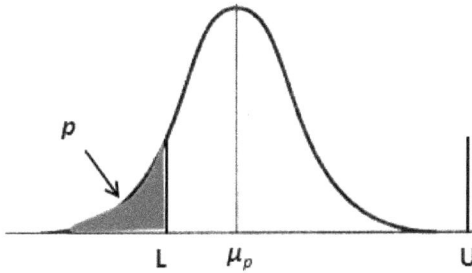

5.6   k-Method: Distance of Sample Mean from Specification Limit

To achieve a certain AQL plan means that we want to test whether p<AQL. This is done by testing whether the process mean is sufficiently far from the specification limits.

Using $\bar{x}$ we compute the quantity $Q_L = \frac{\bar{x}-L}{s}$ and/or $Q_U = \frac{U-\bar{x}}{s}$. Q_L and Q_U measure how far the sample mean is from the specification limit in terms of standard deviations. When σ is unknown, we use the sample standard deviation instead.

Lower Specification Limit

Consider a sampling plan for a lower specification case, with sample size n and acceptance threshold $k = k_L$. We accept the batch if $Q_L > k$ and otherwise reject it. The probability of accepting a batch is given by:

$$OC(p) = P(Q_L > k) = P(\bar{x} > L + \sigma k). \qquad (5.1)$$

Because the measurements follow a normal distribution with mean μ_p and standard deviation σ, the sample mean follows a normal distribution with the same mean and with standard deviation equal to $\sigma/\sqrt{n}$. Computing the probability of acceptance in equation (5.1) requires finding μ_p. The formula for μ_p is given by:

$$\mu_p = L - \sigma z_p \qquad (5.2)$$

where z_p is the pth percentile of the standard normal distribution. z_p can be computed in Excel 2010 using the function NORM.S.INV with parameter p.

Finally, we can compute $OC(p)$ with Excel or an online calculator using normal probabilities with mean μ_p and standard deviation $\sigma/\sqrt{n}$.

To illustrate the computations, consider the organic milk example with the plan n=16, k=2. Let us compute the probability of accepting a batch if the proportion of bottles with less than L=3% fat in the process is p=0.01 and the process standard deviation of fat is 0.25%.

We can compute the process mean by

$$\mu_{0.01} = 3 - 0.25z_{0.01} = 3 - (0.25)(-2.326) = 3.58.$$

In other words, the process mean is set to 3.58% fat, producing 1% of non-conforming milk bottles (with less than 3% fat). The probability of accepting a batch with this quality, is $OC(0.01) = P(\bar{x} > 3 + 0.25 \times 2) = P(\bar{x} > 3.5)$.

Using Excel 2010, we get $OC(0.01)$= 1-NORM.DIST(3.5,3.58,0.25/4,1)= 0.9 (see Figure 5.2 for the NORM.DIST result).

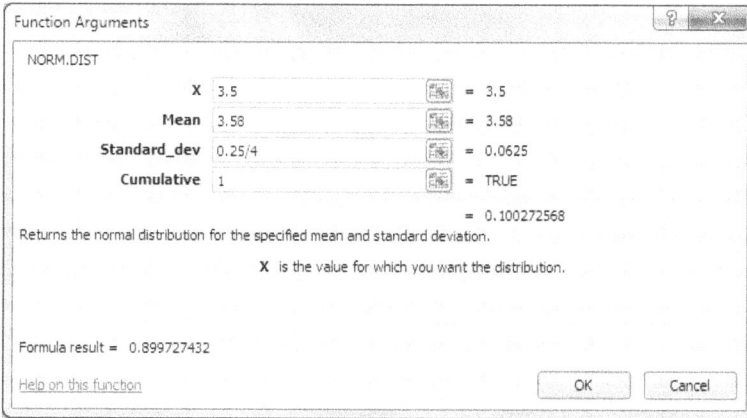

Figure 5.2: Computing the probability of acceptance using Excel 2010

Unknown σ: When the process standard deviation is unknown, we substitute σ with the sample standard deviation s. The computation of $OC(p)$ remains the same except for the need to use the T distribution in place of the normal distribution when the sample size is smaller than n=30 (when n>30 the normal distribution and the T distribution are practically identical).

For example, if in the organic milk example the standard deviation was unknown, we'd use the sample standard deviation s=0.25, and compute $OC(p)$ using the T distribution with $n - 1 = 15$ degrees of freedom:

$$
\begin{aligned}
OC(0.01) &= P(\bar{x} > 3.5) = P\left(T_{15} > \frac{3.5 - 3.58}{0.25/4}\right) \\
&= P(T_{15} > -1.28) = 1 - T.DIST(-1.28, 15, 1) = 0.89
\end{aligned}
$$

Note that this probability of acceptance is slightly lower than in the known σ case, reflecting the extra uncertainty.

Upper Specification Limit

For the case of an upper specification limit, the probability of batch acceptance is given by

$$OC(p) = P(Q_U > k) = P(\bar{x} < U - \sigma k). \qquad (5.3)$$

The process mean formula is given by

$$\mu_p = U - \sigma z_{1-p}. \qquad (5.4)$$

5.7   M-Method: Limit on Proportion of Non-Conforming Items

The proportion of non-conforming items in the process (p) is equal to the probability that a measurement exceeds the specification limit(s). According to the M-Method, the proportion of non-conforming items in the process (p) is estimated from the sample and compared to a threshold M. If M is exceeded, the batch is rejected. Otherwise, it is accepted.

Lower Specification Limit

The proportion of non-conforming items is given by

$$p = P(X < L) \qquad (5.5)$$

where X is assumed to follow a $N(\mu_p, \sigma^2)$ distribution. We can estimate p by computing the normal probability in equation (5.5) with mean $\bar{x}$ and standard deviation σ. This probability can be computed with Excel or an online calculator.

For example, consider the organic milk example. Suppose that after drawing the sample of $n=16$ bottles we obtain an average percentage of fat equal to $\bar{x}=3.5$. We then estimate the percent of non-conforming milk bottles (with less than 3% fat) using the normal distribution with mean 3.5 and standard deviation 0.25 (either known or estimated from the sample) and obtain 0.023 (see Figure 5.3).

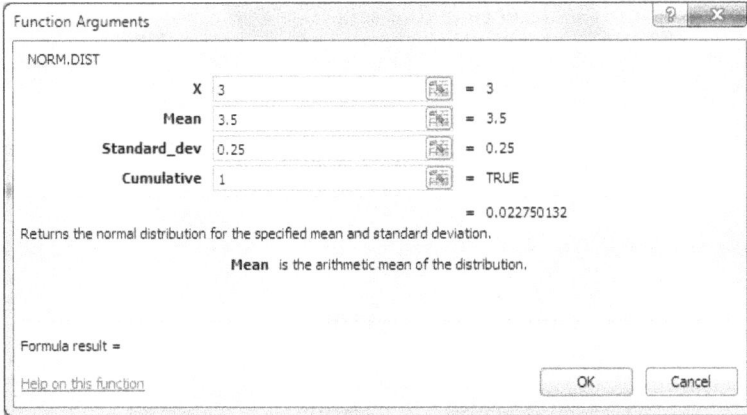

Figure 5.3: Computing estimated proportion of non-conforming items using Excel 2010

While this is a reasonable estimate of p, an improved estimator is achieved by a applying a correction to this number. In particular, the improved estimate ($\hat{p}$) is computed using the formula:

$$\hat{p} = P\left(Z < \sqrt{\frac{n}{n-1}} \times \frac{L - \bar{x}}{s}\right) = P\left(Z > \sqrt{\frac{n}{n-1}} Q_L\right) \quad (5.6)$$

where $Z \sim N(0,1)$. Note that the second expression links between the k-Method and the M-Method.

For the organic milk example the improved estimate yields

$$\hat{p} = P\left(Z < \sqrt{\frac{16}{15}} \times \frac{3 - 3.5}{0.25}\right) = P(Z < -2.066) = 0.019.$$

Finally, we compare $\hat{p}$ with the threshold number M in order to decide whether to accept or reject the batch. In the case of a single specification limit (L or U), M is related to k (from the k-Method) via the formula

$$M = P\left(Z > \sqrt{\frac{n}{n-1}} k\right). \quad (5.7)$$

Upper Specification Limit

Estimating the proportion of non-conforming items in the process when there is an upper specification limit is very similar to

the lower specification limit case. The only difference is that the
formula for p is given by

$$p = P(X > U) \tag{5.8}$$

This probability is estimated by computing a normal probability
with mean $\bar{x}$ and standard deviation σ (or the sample standard
deviation (s) when σ is unknown):

$$\hat{p} = P\left(Z > \sqrt{\frac{n}{n-1}} \times \frac{U - \bar{x}}{s}\right) = P\left(Z > \sqrt{\frac{n}{n-1}} Q_U\right). \tag{5.9}$$

Double Specification Limit

In applications with both lower and upper specification limits, a
batch is accepted only if $1 - P(L < X < U)$ does not exceed a threshold
M. The formal procedure is to estimate $P_L = P(X < L)$ and $P_U =
P(X > U)$ using the formulas in the single-specification cases, and
to compare their sum to a threshold M that is the sum of the
upper and lower thresholds.

5.8   ANSI/ASQC Z1.9 Tables (Military Standard 414)

The military standard for variable (continuous) data is Mil-Std-
414, with the last version being cancelled in 1999. The equivalent
civilian counterpart by the International Standardization Organization (ISO) is called *ISO 3951*. The equivalent standard by the
American National Standard Institute (ANSI) is *ANSI/ASQC Z1.9*
(released in 1980). The equivalent British standard is *BS 6002*.

Unlike the equivalence between Mil-Std-105 and its civilian
counterparts, the civilian counterparts of Mil-Std-414 were revised in 1990 to make them as similar in structure to ANSI/ASQC
Z1.4. In particular, the civilian tables differ from Mil-Std-414 in
the following ways:

- Lot size ranges in the code letter table are identical to those in
  ANSI/ASQC Z1.4 (Figure 2.9)

- Mil-Std-414 code letters J,L were removed. Mil-Std-414 code
  letters K,M,N,O,P,Q were converted in ANSI/ASQC Z1.9 to
  J,K,L,M,N,P

- Inspection levels I-IV were renamed to S-3, S-4, I,II, III to match ANSI/ASQC Z1.4 levels

- Mil-Std-414 AQL levels 0.04%, 0.065% and 15% were eliminated

- The switching rules were replaced with those from ANSI/ASQC Z1.4

Finally, there are slight differences between ANSI/ASQC Z1.9 and ISO 3951 in terms of the available tables. Many companies and organizations use the ANSI/ASQC Z1.9 (or equivalent) standard because it is required for certification.

Using the ANSI/ASQC Z1.9 or Mil-Std-414 tables is slightly more involved than the standards for attribute data, as there are multiple tables to look up for obtaining a sampling plan and an accept/reject decision. We start by describing the different tables and their use, and then describe the much simpler (and less error-prone) online application at SQCOnline.com. We will focus on Mil-Std-414 and ANSI/ASQC Z1.9.

Design of Tables

The tables give inspection plans for a given batch size (N) and Acceptable Quality Level (AQL). These plans are therefore producer-based, assuming a long-term relationship between the producer and consumer. The plans guarantee a producer's risk (α) between 1% and 10%. To control the consumer's risk, there are various inspection levels.

Table Usage

The ISO 3951 and ANSI/ASQC Z1.9 standard is available for purchase in hardcopy from various organizations. Mil-Std-414 tables can be downloaded in PDF format from www.sqconline. com/download.

Code Letter   Finding a sampling plan starts with specifying the *batch size* (lot size) and the *inspection level*. Figure 5.4 shows the table where the lot size and inspection level determine a code letter. The default inspection level in ANSI/ASQC Z1.9 is level II, as in the ANSI/ASQC Z1.4 standard for attributes. Similarly, Inspection levels S3,S4, and I lead to more lenient inspection (which means higher consumer risk), and inspection level III leads to more stringent inspection (larger samples; steeper OC curve). Note also that the Lot Size ranges in ANSI/ASQC Z1.9 are identical to ANSI/ASQC Z1.4.

In Mil-Std-414 the default inspection level is IV. This is equivalent to level II in ANSI/ASQC Z1.9.

Another initial table to examine is the table mapping ranges of AQLs to the value of AQL to use in the sampling plan table. Figure 5.5 shows the Mil-Std-414 table. The greyed out rows correspond to the rows removed in ANSI/ASQC Z1.9.

Let us use the organic milk example to illustrate the use of the tables. Recall that batch sizes are 5,000 bottles and an AQL=1% is required. Using the ANSI/ASQC Z1.9 code letter table with inspection level II we obtain code letter L. Using the AQL range table, we obtain AQL=1%. Let us assume that σ is unknown.

Inspection Types   As in ANSI/ASQC Z1.4, there are three types of inspection: normal, tightened, and reduced. The same guidelines and switching rules as in ANSI/ASQC Z1.4 apply (see Section 3.5)

Sampling Plan Tables   Once the inspection level is selected, the user must choose between the k-Method (only for single-specification limit cases) and the M-Method. Also, one must choose between tables for known σ, unknown σ using the sample standard deviation, or unknown σ using the sample range. Figure 5.6, for instance, displays a Mil-Std-414 k-Method table for normal and tightened inspection, unknown σ, using the sample standard deviation. The greyed out rows and columns (signifying cancelled parts) and the new code letters on the left show the revisions that now constitute ANSI/ASQC Z1.9.

Continuing our organic milk example, a normal inspection plan using code letter L gives the plan $n=75$, $k=1.98$. This means that after sampling 75 bottles of milk, we should compute the

Lot Size			Inspection Levels				
			Special S3	S4	General I	II	III
2	to	8	B	B	B	B	C
9	to	15	B	B	B	B	D
16	to	25	B	B	B	C	E
26	to	50	B	B	C	D	F
51	to	90	B	B	D	E	G
91	to	150	B	C	E	F	H
151	to	280	B	D	F	G	I
281	to	400	C	E	G	H	J
401	to	500	C	E	G	I	J
501	to	1,200	D	F	H	J	K
1,201	to	3,200	E	G	I	K	L
3,201	to	10,000	F	H	J	L	M
10,001	to	35,000	G	I	K	M	N
35,001	to	150,000	H	J	L	N	P
150,001	to	500,000	H	K	M	P	P
500,001	and	over	H	K	N	P	P

LOT SIZE			INSPECTION LEVELS				
			I	II	III	IV	V
3	to	8	B	B	B	B	C
9	to	15	B	B	B	B	D
16	to	25	B	B	B	C	E
26	to	40	B	B	B	D	F
41	to	65	B	B	C	E	G
66	to	110	B	B	D	F	H
111	to	180	B	C	E	G	I
181	to	300	B	D	F	H	J
301	to	500	C	E	G	I	K
501	to	800	D	F	H	J	L
801	to	1,300	E	G	I	K	L
1,301	to	3,200	F	H	J	L	M
3,201	to	8,000	G	I	L	M	N
8,001	to	22,000	H	J	M	N	O
22,001	to	110,000	I	K	N	O	P
110,001	to	550,000	I	K	O	P	Q
550,001	and	over	I	K	P	Q	Q

Figure 5.4: Code letter table
for ANSI/ASQC Z1.9 (left)
and Mil-Std-414 table (right).

For specified AQL values falling within these ranges	Use this AQL value
———— to 0.049	0.04
0.050 to 0.069	0.065
0.070 to 0.109	0.10
0.110 to 0.164	0.15
0.165 to 0.279	0.25
0.280 to 0.439	0.40
0.440 to 0.699	0.65
0.700 to 1.09	1.0
1.10 to 1.64	1.5
1.65 to 2.79	2.5
2.80 to 4.39	4.0
4.40 to 6.99	6.5
7.00 to 10.9	10.0
11.00 to 16.4	15.0

Figure 5.5: Mil-Std414 table mapping range of AQL to the AQL value to use in the tables. ANSI/ASQC Z1.9 excludes the first and last two rows

Figure 5.6: Mil-Std-414 Table — k values (σ unknown, standard deviation method)

ACCEPTABLE QUALITY LEVELS (NORMAL INSPECTION) — top AQL label of each column.
ACCEPTABLE QUALITY LEVELS (TIGHTENED INSPECTION) — bottom (staggered) AQL label of each column.

Z1.9 new code	Sample Size Code Letter	Sample Size	AQL 0.04	0.065	0.10	0.15	0.25	0.40	0.65	1.00	1.50	2.50	4.00	6.50	10.0	15.0
(tightened AQL)			—	0.065	0.10	0.15	0.25	0.40	0.65	1.00	1.50	2.50	4.00	6.50	10.00	15.0
	B	3	↓	↓	↓	↓	↓	↓	↓	↓	↓	1.12	0.958	0.765	0.566	0.341
	C	4	↓	↓	↓	↓	↓	↓	↓	1.45	1.34	1.17	1.01	0.814	0.617	0.393
B	D	5	↓	↓	↓	↓	↓	↓	1.65	1.53	1.40	1.24	1.07	0.874	0.675	0.455
C	E	7	↓	↓	↓	↓	2.00	1.88	1.75	1.62	1.50	1.33	1.15	0.955	0.755	0.536
D	F	10	↓	↓	↓	2.24	2.11	1.98	1.84	1.72	1.58	1.41	1.23	1.03	0.828	0.611
E	G	15	2.64	2.53	2.42	2.32	2.20	2.06	1.91	1.79	1.65	1.47	1.30	1.09	0.886	0.664
F	H	20	2.69	2.58	2.47	2.36	2.24	2.11	1.96	1.82	1.69	1.51	1.33	1.12	0.917	0.695
G	I	25	2.72	2.61	2.50	2.40	2.26	2.14	1.98	1.85	1.72	1.53	1.35	1.14	0.936	0.712
H	J	30	2.73	2.61	2.51	2.41	2.28	2.15	2.00	1.86	1.73	1.55	1.36	1.15	0.946	0.723
I	K	35	2.77	2.65	2.54	2.45	2.31	2.18	2.03	1.89	1.76	1.57	1.39	1.18	0.969	0.745
J	L	40	2.77	2.66	2.55	2.44	2.31	2.18	2.03	1.89	1.76	1.58	1.39	1.18	0.971	0.746
K	M	50	2.83	2.71	2.60	2.50	2.35	2.22	2.08	1.93	1.80	1.61	1.42	1.21	1.00	0.774
L	N	75	2.90	2.77	2.66	2.55	2.41	2.27	2.12	1.98	1.84	1.65	1.46	1.24	1.03	0.804
M	O	100	2.92	2.80	2.69	2.58	2.43	2.29	2.14	2.00	1.86	1.67	1.49	1.26	1.05	0.819
N	P	150	2.96	2.84	2.73	2.61	2.47	2.33	2.18	2.03	1.89	1.70	1.51	1.29	1.07	0.841
P	Q	200	2.97	2.85	2.73	2.62	2.47	2.33	2.18	2.04	1.89	1.70	1.51	1.29	1.07	0.845

Figure 5.6: Mil-Std-414 Table for normal and tightened inspection, single specification limit, k-Method, σ unknown, (sample) standard deviation method. ANSI/ASQC Z1.9 excludes the greyed out rows and columns. Their new code letters are marked to the left

average and standard deviation of the fat percentage, and then compare $Q_L = \frac{\bar{x}-3}{s}$ to k=1.98. If $Q_L > k$, we accept the batch. For example, if $\bar{x} = 3.5$ and $s = 0.25$, then $Q_L = 2$ and we accept the batch. Or equivalently, we compare $\bar{x}$ to 3+1.98s. If it exceeds this limit, we accept the batch (indicating that the process mean is likely sufficiently far from L).

The corresponding tightened inspection plan, using the AQL levels at the bottom row, yields n=75, k=2.12. This more stringent plan means that we require a larger distance between the sample mean and the lower specification limit L=3 to accept the batch. With the value Q_L=2 that we obtained above, we would reject the batch under tightened inspection.

Let us now see an example of the M-Method procedure. Figure 5.7 gives the same table as Figure 5.6, except that it uses the M-Method. Using code letter L we obtain the normal inspection plan n=75, M=2.29%. This means that after drawing a sample of 75 bottles and measuring the fat percentage, we estimate the proportion non-conforming $\hat{p}$ and compare it to M.

In Section 5.7 we gave a formula for $\hat{p}$. In ANSI/ASQC Z1.9, this formula is used only in the σ known case. Otherwise, a special conversion table gives the value of $\hat{p}$ for the corresponding value of Q_L (or Q_U for an upper specification limit).

Figure 5.8 shows part of the conversion table when using the sample standard deviation. Given the sample size and Q_L we obtain the value for $\hat{p}$. For the organic milk example, using n=75 and Q_L=2, we get $\hat{p}$=2.16%. This estimated percent of non-conforming bottles does not exceed the threshold M=2.29% and therefore we accept the batch.

Online Plans

Obtaining sampling plans using the Mil-Std-414 standard is somewhat confusing and error-prone. Finding the code letter, then the appropriate inspection table, reading the plan parameters (using the arrows and reading footnotes), obtaining values of $\hat{p}$, and understanding the resulting decision rules can be taxing to users and can easily lead to mistakes.

A more user-friendly and error-proof solution is to use the

	SAMPLE SIZE CODE LETTER	SAMPLE SIZE	ACCEPTABLE QUALITY LEVELS (NORMAL INSPECTION)													
			0.04	0.065	0.10	0.15	0.25	0.40	0.65	1.00	1.50	2.50	4.00	6.50	10.0	15.0
			M	M	M	M	M	M	M	M	M	M	M	M	M	M
	B	3	↓	↓	↓	↓	↓	↓	↓	↓	↓	7.59	18.86	26.94	33.69	40.47
	C	4	↓	↓	↓	↓	↓	↓	↓	1.53	5.50	10.92	16.45	22.86	29.45	36.90
	D	5	↓	↓	↓	↓	↓	↓	1.33	3.32	5.83	9.80	14.39	20.19	26.56	33.99
	E	7	↓	↓	↓	↓	0.422	1.06	2.14	3.55	5.35	8.40	12.20	17.35	23.29	30.50
	F	10	↓	↓	↓	0.349	0.716	1.30	2.17	3.26	4.77	7.29	10.54	15.17	20.74	27.57
	G	15	0.099	0.186	0.312	0.503	0.818	1.31	2.11	3.05	4.31	6.56	9.46	13.71	18.94	25.61
	H	20	0.135	0.228	0.365	0.544	0.846	1.29	2.05	2.95	4.09	6.17	8.92	12.99	18.03	24.53
	I	25	0.155	0.250	0.380	0.551	0.877	1.29	2.00	2.86	3.97	5.97	8.63	12.57	17.51	23.97
	J	30	0.179	0.280	0.413	0.581	0.879	1.29	1.98	2.83	3.91	5.86	8.47	12.36	17.24	23.58
J	K	35	0.170	0.264	0.388	0.535	0.847	1.23	1.87	2.68	3.70	5.57	8.10	11.87	16.65	22.91
	L	40	0.179	0.275	0.401	0.566	0.873	1.26	1.88	2.71	3.72	5.58	8.09	11.85	16.61	22.86
K	M	50	0.163	0.250	0.363	0.503	0.789	1.17	1.71	2.49	3.45	5.20	7.61	11.23	15.87	22.00
L	N	75	0.147	0.228	0.330	0.467	0.720	1.07	1.60	2.29	3.20	4.87	7.15	10.63	15.13	21.11
M	O	100	0.145	0.220	0.317	0.447	0.689	1.02	1.53	2.20	3.07	4.69	6.91	10.32	14.75	20.66
N	P	150	0.134	0.203	0.293	0.413	0.638	0.949	1.43	2.05	2.89	4.43	6.57	9.88	14.20	20.02
P	Q	200	0.135	0.204	0.294	0.414	0.637	0.945	1.42	2.04	2.87	4.40	6.53	9.81	14.12	19.92
			0.065	0.10	0.15	0.25	0.40	0.65	1.00	1.50	2.50	4.00	6.50	10.00	15.00	
			ACCEPTABLE QUALITY LEVELS (TIGHTENED INSPECTION)													

Figure 5.7: Mil-Std-414 Table for normal and tightened inspection, single specification limit, M-Method, σ unknown, (sample) standard deviation method. ANSI/ASQC Z1.9 excludes the greyed out rows and columns. Their new code letters are marked to the left

Q_U or Q_L	Sample Size															
	3	4	5	7	10	15	20	25	30	35	40	50	75	100	150	200
2.00	0.00	0.00	0.00	0.43	1.17	1.62	1.81	1.91	1.98	2.03	2.06	2.10	2.16	2.19	2.22	2.23
2.01	0.00	0.00	0.00	0.39	1.12	1.57	1.76	1.86	1.93	1.97	2.01	2.05	2.11	2.14	2.17	2.18
2.02	0.00	0.00	0.00	0.34	1.07	1.52	1.71	1.81	1.87	1.92	1.95	2.00	2.06	2.09	2.11	2.13
2.03	0.00	0.00	0.00	0.32	1.03	1.47	1.66	1.76	1.82	1.87	1.90	1.95	2.01	2.04	2.06	2.08
2.04	0.00	0.00	0.00	0.29	0.98	1.42	1.61	1.71	1.77	1.82	1.85	1.90	1.96	1.99	2.01	2.06
2.05	0.00	0.00	0.00	0.26	0.94	1.37	1.56	1.66	1.73	1.77	1.80	1.85	1.91	1.94	1.96	1.98

Figure 5.8: ANSI/ASQC Z1.4 Table for estimating the lot percent non-conforming for unknown σ using the sample standard deviation

website www.sqconline.com. The current version implements Mil-Std-414 and ANSI/ASQC Z1.9. The user is first requested to choose between known/unknown σ and to enter four process parameters using pull-down menus: batch size, AQL, inspection level (I-V for Mil-Std-414 or S-3, S-4 I-III for ANSI/ASQC Z1.9), and type of inspection (normal/reduced/tightened). If the user is not sure about the meaning of a certain parameter, they can click on *more info* to obtain more details (see Figure 5.9, showing the input page for Mil-Std414).

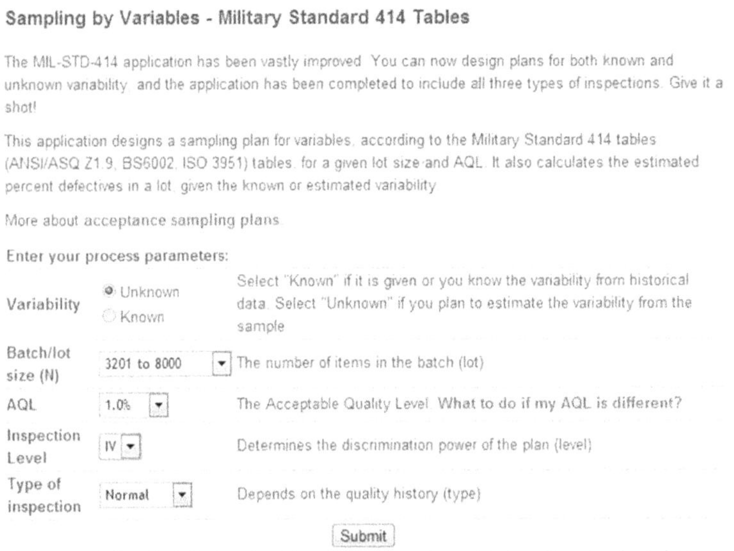

Sampling by Variables - Military Standard 414 Tables

The MIL-STD-414 application has been vastly improved. You can now design plans for both known and unknown variability, and the application has been completed to include all three types of inspections. Give it a shot!

This application designs a sampling plan for variables, according to the Military Standard 414 tables (ANSI/ASQ Z1.9, BS6002, ISO 3951) tables, for a given lot size and AQL. It also calculates the estimated percent defectives in a lot, given the known or estimated variability.

More about acceptance sampling plans

Enter your process parameters:

Variability	● Unknown ○ Known	Select "Known" if it is given or you know the variability from historical data. Select "Unknown" if you plan to estimate the variability from the sample
Batch/lot size (N)	3201 to 8000 ▼	The number of items in the batch (lot)
AQL	1.0% ▼	The Acceptable Quality Level. What to do if my AQL is different?
Inspection Level	IV ▼	Determines the discrimination power of the plan (level)
Type of inspection	Normal ▼	Depends on the quality history (type)

Submit

Figure 5.9: Input screen on SQCOnline.com for Mil-Std-414

The output page (Figure 5.10) gives the M-Method sampling plan in terms of the needed sample size and the decision rule for acceptance. For example, entering the parameters for the organic milk example into the Mil-Std-414 calculator (N=5000, AQL=1%, normal inspection, level IV), we get the plan $n=50$, $M=2.49\%$.

The user is then prompted to enter the specification limit(s) and sample results to obtain $\hat{p}$. For our example, we enter $\bar{x}=3.5, s=0.25, L=3$. We then get the final screen (Figure 5.11) with $\hat{p}=2.10\%$ and the resulting decision to accept the batch.

Sampling by Variables

Military Standard 414 (ANSI/ASQC Z1.9) Tables

For a lot of 3201 to 8000 items. and AQL= 1.0%   with inspection level IV the **Normal** inspection plan is

Sample 50* items.

If the estimated percent of non-conforming (defective) items is
2.49% or less --> accept the lot
Otherwise. reject it

To estimate the percent of non-conforming items in your process take a sample of size 50 and enter the values into the next table

To estimate your process % non-conforming (defectives) enter:

Sample Average (x̄):	3.5	The average of the 50 measurements
Process Standard Deviation (s):	0.25	The standard deviation of the 50 measurements
Lower Specification Limit:	3	The smallest value for your measurement that is considered acceptable. Leave blank if there is no lower limit
Upper Specification Limit:		The largest value for your measurement that is considered acceptable. Leave blank if there is no upper limit

Submit

*Note If the sample size exceeds the lot size. apply 100% sampling of the lot

Figure 5.10: Output screen from SQCOnline.com, giving the M-method sampling plan. The next screen helps compute $\hat{p}$

5.9   Sampling Plans for Specified Producer and Consumer Risks

As in the case of sampling by attributes, the military standard and its civilian counterparts are based on assuring a producer's risk of 1%-10% for the required AQL.

In some cases the producer and consumer require a plan that satisfies two criteria: it assures that AQL-quality batches will be accepted with high probability (low α) and LTDP-quality batches will be rejected with high probability (low β). The plan is intended to be used for a long-term relationship, thereby protecting both the producer and the consumer's interests.

In the following we give formulas for obtaining a plan (n and

Sampling by Variables

Military Standard 414 (ANSI/ASQC Z1.9) Tables

Here is a summary of your process parameters, the sampling plan, and the sample statistics:

Information Summary

Process Parameters	Lot size = 3201 to 8000 AQL = 1.0% Lower Spec. Limit (L) = 3 Upper Spec. Limit (U) = Standard Dev. (s) = 0.25 Variability = unknown
Sampling Plan	Sample size (n) = 50 Non-conforming limit (M) = 2.49% Inspection Type = Normal Inspection Inspection Level = IV
Sample statistics	Sample average (x) = 3.5

Based on these values, the estimated proportion of non-conforming (defective) items in your process is 2.12%.

Since this percent is not higher than 2.49%, the lot should be accepted.

The following table details the calculations involved in computing the above results, as detailed in MIL-STD-414
It is provided for reference only.

Calculation steps for unknown variability

Information needed	Value	Explanation
Sample size code letter	M	See Table A-2 in MIL-STD-414
Lower Quality Index: Q_L	2.00	$(\bar{x}\text{-L})/\sigma$
Est. of lot % defective below Lower Spec: P_L	2.12%	See Table B-5 in MIL-STD-414
Max. allowable % defective: M	2.49%	See Table B-3/B-4 in MIL-STD-414

Figure 5.11: Final output screen from SQCOnline.com displaying $\hat{p}$ and the accept/reject decision. The lower table shows the interim calculations

k) for the case of a single specification limit.

For a lower specification limit, the two equations to solve in order to obtain a sampling plan with specified α and β are:

$$
\begin{aligned}
1-\alpha &= OC(AQL) = P(Q_L > k) \\
&= P(\bar{x} > L + \sigma k \mid \mu_{AQL}) \quad\quad (5.10) \\
\beta &= OC(LTPD) = P(Q_L > k) \\
&= P(\bar{x} > L + \sigma k \mid \mu_{LTPD}) \quad\quad (5.11)
\end{aligned}
$$

A similar set of equations can be written for an upper specification limit. These two probabilities are calculated using the normal distribution with mean μ_{AQL} and μ_{LTPD} respectively, and standard deviation $\sigma/\sqrt{n}$. For the lower or upper specification limit case, we can solve for n and k and get the following formula for k:

$$
k = \frac{Z_{1-\alpha}Z_{1-LTPD} + Z_{1-\beta}Z_{1-AQL}}{Z_{1-\alpha} + Z_{1-\beta}} \quad\quad (5.12)
$$

The notation $Z_{1-\alpha}$ means the $1-\alpha$ percentile of a standard normal distribution (NORM.S.INV($1\text{-}\alpha$)). The sample size depends on whether σ is known or not. For known σ, the sample size is given by

$$n = \left(\frac{Z_{1-\alpha} + Z_{1-\beta}}{Z_{1-AQL} - Z_{1-LTPD}} \right)^2 \qquad (5.13)$$

and for unknown σ the sample size is larger by a factor of $1 + \frac{k^2}{2}$, yielding:

$$n = \left(1 + \frac{k^2}{2} \right) \left(\frac{Z_{1-\alpha} + Z_{1-\beta}}{Z_{1-AQL} - Z_{1-LTPD}} \right)^2 . \qquad (5.14)$$

For example, let us find the inspection plan for a single specification limit case where the producer's risk at AQL=0.01 should be α=0.05 and the consumer's risk at LTPD=0.06 should be β=0.10. If σ is known, we get the plan

$$k = \frac{(1.64)(1.55) + (1.28)(2.33)}{1.64 + 1.28} = 1.89$$

and n=15 (after rounding the number 14.38). If σ is unknown, we use the same k=1.89 with sample size n=30 (14.38 multiplied by the factor $1 + \frac{1.432}{2}$).

5.10   Problems

1. *Bee Healthy* is a coop of honey collectors in India. The coop regularly inspect batches of their honey jars to assure that quality is up to the standards of an organic certifying agency. A honey jar is conforming if it contains at least 1700ml. Each batch consists of 1,000 jars. From experience it is known that the amount of honey in a jar is approximately normally distributed. The coop would like to employ a sampling plan that will guarantee for AQL=1% a producer's risk of α=5%.

   (a) If 5% of the jars contain less than 1700ml of honey, what is the average volume of honey per jar?

   (b) Find an inspection plan using the ANSI/ASQC Z1.9 standard, using the k-Method. Use normal inspection, level II. Report the plan parameters and explain the procedure and decision rule in terms of the sample mean.

(c) Find an inspection plan using the ANSI/ASQC Z1.9 standard (or SQCOnline.com), using the M-Method. Use normal inspection, level II. Report the plan parameters and explain the procedure and decision rule in words.

(d) Find the corresponding Mil-Std-414 plan using SQCOnline.com. Is there a difference in the resulting plan?

(e) A sample of size n was taken according to the plan in (b). The average of the honey volume was found to be 1910ml, with a standard deviation of 50ml. What should be the decision regarding the batch of 1,000 jars?

(f) Using the estimated $s=50$, what is the probability of accepting a batch if the quality is 1%? Does this satisfy the requirement $\alpha=5\%$?

(g) How can the coop achieve the requirement $\alpha=5\%$ for AQL=1%? Mention two options.

2. *Save Trees* is a vendor of recycled paper. It receives batches of 2500 pages of recycled paper from a recycling facility every week. The weight (called grammage) of each sheet of paper should be at least 4.9g. The required AQL is 1%. Assume that the process standard deviation is unknown.

(a) Find an inspection plan using the ANSI/ASQC Z1.9 standard M-method, using normal inspection, level II. Report the plan parameters and explain the procedure and decision rule in words.

(b) Find a similar plan with tightened inspection. How does it differ from the normal inspection plan in (a)?

(c) Find the equivalent normal plan in Mil-Std-414. Report the parameters. Is the plan identical to the one obtained from ANSI/ASQC Z1.9?

(d) A sample of size n was taken according to the plan in (c). The sample average was 5.05g and the standard deviation was 0.05g. Would the batch be accepted according to the Mil-Std-414 plan?

(e) What is the underlying assumption which must be met in order for the inspection plan to yield valid decisions?

(f) *Save Trees* and the recycling facility have signed a long-term contract where the recycling facility will provide paper to *Save Trees* on a weekly basis for 3 years. They have agreed to implement an inspection plan that protects both parties. In particular, the inspection scheme should accept of batches of quality AQL=1% with probability 0.95, and should reject batches of quality LTPD=10% with probability 0.98. Using the normal approximation, find such a plan and report its parameters.

3. *Bite* is a manufacturer of plastic materials for dental care. One of their products is a plastic membrane that includes an active ingredient called Acetyltribultyl Citrate (ATBC). The amount of ATBC must range between 14.0%- 15.3% in order to meet specifications. The manufacturer inspects the membranes two years after manufacturing to assure that the product meets a required quality level of AQL=6.5%. Batches are typically of size $N=500$ membranes.

(a) Find an inspection plan using the ANSI/ASQC Z1.9 standard, using normal inspection, level II. Report the plan parameters and explain the procedure and decision rule in words.

(b) Find an inspection plan using the ANSI/ASQC Z1.4 standard (for plans by attributes), using normal inspection, level II. Report the plan parameters and explain the procedure and decision rule in words.

(c) What is the advantage of treating the active ingredient inspection as a pass/fail classification rather than an exact measurement? What is the disadvantage?

6 *Continuous Sampling*

In previous chapters we looked at *lot-by-lot* acceptance sampling plans, which are sampling inspection plans for batches. Batches arise naturally in many applications: when raw materials are packed or shipped in batches, when services are delivered on a daily, weekly, or other basis, or in general, when items are grouped in some natural way.

In some cases, the notion of batches is inapplicable, impractical or artificial, and instead units are produced individually in a continuous flow, in assembly-line fashion. One example is the process of data entry, where a flow of data arrives and is typed by a technician. Another example is in the manufacturing of complex or expensive items such as automobiles or aircraft engines, where waiting for a *batch* is impractical. Another common environment where lot-by-lot sampling is impractical is in customized or on-demand manufacturing. Examples include Dell computers and Nike's custom sneakers (NIKEid), where consumers customize their wanted products and these are then produced and immediately shipped out. It is clearly impossible to create lots in such an environment.

In 1943, Harold F. Dodge developed a sampling plan, which is now termed *Continuous Sampling Plan 1* (CSP-1) aimed at providing assurance that the long run percentage of defective units in the accepted product will be held down to a prescribed limit. Later, further modifications were proposed and termed CSP-2 and CSP-3. In the early fifties, the American Army Ordinance Corps became interested in the application of CSP to the inspection of rounds of ammunition components. The two production conditions that favored continuous sampling over lot-by-lot sampling were the continuous flow of ammunition items from one stage of assembly to another, and the limited storage space for explosive materials at an inspection station. The American Navy and Air Force soon followed and started using continuous sampling plans for inspecting items such as aircraft engines and transmitters.

Continuous sampling plans consist of two phases: a *screening phase* where all units are inspected (100% inspection), and a *sampling phase* where a fraction f ($0 < f < 1$) of the units is inspected. Operationally, this means that the following requirements must be met for employing CSP (as dictated by the Military Standard 1235):

- Possibility for rapid 100% inspection when necessary

- Relative ease and speed of inspection

- Inspection is non-destructive

We note that continuous sampling can be used for two purposes: for *product screening* in the sense of separating units into acceptable and unacceptable grade, and for *process troubleshooting* in terms of monitoring quality (when inspection rates rapidly increase, this indicates a deterioration in quality). In this chapter we describe the most popular plan CSP-1, as well as additional plans such as CSP-2. All these plans are based pass/fail classifications of units. Plans can be carried out using rectifying inspection, where non-conforming units are rectified or replaced, or using non-rectifying inspection where non-conforming units are discarded. In the following sections, we discuss the design and implementation of these plans, their performance, the use of Military Standard 1235, and extensions to inspecting continuous flows of batches (skip-lot sampling).

6.1   Continuous Sampling Plan 1 (CSP-1)

Procedure

Continuous Sampling Plan 1 is the simplest of the continuous sampling schemes. It has two phases and includes two parameters: the *clearance number* (i) and the *sampling fraction* (f). Inspection alternates between two phases (see Figure 6.1):

1. Inspection starts with a *screening phase*, where each item is inspected. The screening phase continuous until i consecutive conforming units are found.

2. Then, inspection moves to the *sampling phase*, where a fraction f of the units is inspected. When the first non-conforming unit is encountered, the screening phase resumes, and so-forth.

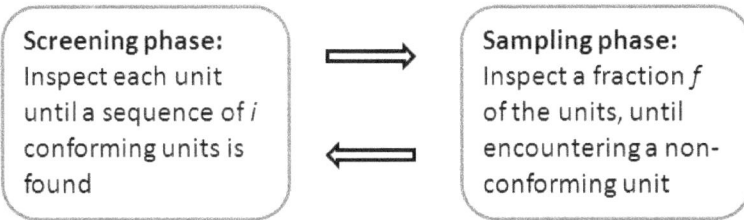

Screening phase:
Inspect each unit
until a sequence of i
conforming units is
found

Sampling phase:
Inspect a fraction f
of the units, until
encountering a non-
conforming unit

Figure 6.1: Schematic of the CSP-1 procedure. Inspection can be rectifying or non-rectifying

While the easiest procedure for sampling in the sampling phase is to sample every $1/f$th unit, called *systematic sampling*, alternatives that are statistically equivalent are *probabilistic sampling*, where each unit is sampled with probability f, and *block-random sampling*, where a unit is chosen at random from a group of $1/f$ consecutive items. The last two procedures are more complicated operationally, but when the process experiences a cyclical change in quality with cycle length $1/f$, then systematic sampling will be at a disadvantage.

Dodge's CSP-1 is a rectifying scheme, where every non-conforming unit that is encountered is rectified or replaced. We start by describing the rectifying scheme and later on consider also a non-rectifying scheme.

Performance Measures and Choosing i and f

Average Outgoing Quality (AOQ)   The main criterion for choosing the parameters i and f is the *Average Outgoing Quality Limit* (AOQL) required by the consumer. Because rectifying inspection is used, the outgoing quality is usually better than the incoming quality. In general, the outgoing quality is improved by increasing the clearance number i, thereby inspecting more units under 100% inspection. The outgoing quality is also improved as we increase the sampling fraction f, such that more units are inspected during the sampling phase.

The choice of i and f influences the length of each of the two phases in terms of the number of units inspected under screening and sampling. In particular, the average number of units exiting the screening phase and sampling phase are given by the formulas

$$E_{Screening} = \frac{1 - (1 - p)^i}{p(1 - p)^i}$$

$$E_{Sampling} = \frac{1}{pf} \tag{6.1}$$

Note that outgoing non-conforming units result only from the uninspected units during the sampling phase. Hence, the average outgoing quality of a process with incoming quality p, when rectifying inspection is employed, is given by the formula

$$AOQ(p) = \frac{p(1 - f)E_{Sampling}}{E_{Screening} + E_{Sampling}} = \frac{p(1 - p)^i}{(1 - p)^i + \frac{f}{1-f}}. \tag{6.2}$$

The consumer wants to assure that regardless of the incoming quality p, the worst-case AOQ will be below a certain level. Hence, the consumer wants to minimize AOQL = $max_p(AOQ(p))$. However, there usually exist multiple pairs (i, f) that satisfy a given AOQL. For example, consider the two rectifying CSP-1 schemes $i=7$, $f=1/3$ and $i=9$, $f=1/4$ shown in Figure 6.2. Their AOQ curves are very similar and both have a maximum of approximately AOQL=0.06. Which plan is better? More generally, how to choose a CSP-1 plan among a set of plans that achieve the same AOQL?

Average Fraction Inspection (AFI)   To choose between plans that guarantee the same AOQL, we consider the producer's costs. The most common measure is the expected proportion of inspected items called *Average Fraction Inspected* (AFI). The AFI

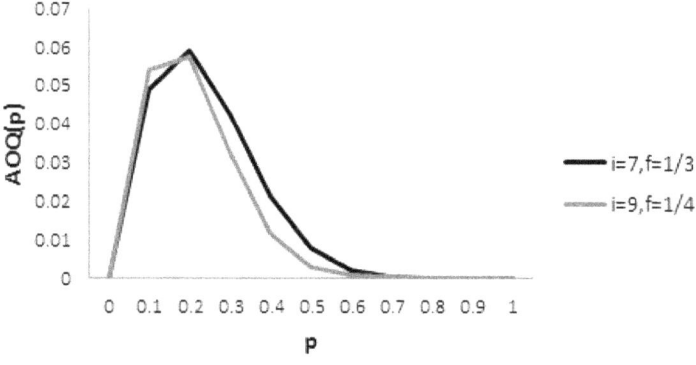

measures the proportion of units inspected per cycle of screening plus sampling phases. Recall that during screening all units are inspected and during sampling only a fraction f is inspected. The formula, for rectifying inspection, is therefore given by

$$AFI(p) = \frac{E_{Screening} + fE_{Sampling}}{E_{Screening} + E_{Sampling}} = \frac{f}{(f + (1-f)(1-p)^i}. \quad (6.3)$$

In general, the relationship between $AOQ(p)$ and $AFI(p)$ is simple: if we inspect an average of $AFI(p)$ units, then that entire portion is free of non-conforming items, while the remaining uninspected portion $1 - AFI(p)$ contains a proportion p of non-conforming items. Hence $AOQ(p) = p(1 - AFI(p))$.

$AFI(p)$ is commonly used to choose between plans that have the same AOQL. For the two CSP-1 plans shown in Figure 6.2 we can compute the AFI for a particular quality level p. Or better, we plot the AFI curves for the two plans in Figure 6.3. We can see that for incoming quality below p=0.2 the plan i=9,f=1/4 has lower $AFI(p)$, whereas for quality worse than p=0.2 the plan i=7, f=1/3 has lower expected fraction inspection.

Considering Additional Costs   AFI considers only the cost of per-unit inspection. In some cases there are additional costs such as the cost of switching between the sampling and screening

AFI Curve for two CSP-1 Schemes

Figure 6.3: AFI curves for the two CSP-1 schemes from Figure 6.2

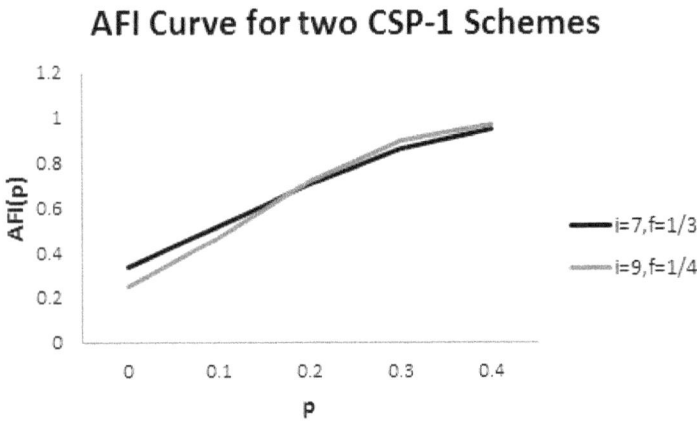

phases. For example, consider a production line that produces one unit per hour, and is operating at a quality of p=0.1. The cost of inspecting an item is \$1 and the cost of switching from sampling to screening or vice versa is \$50. For the two CSP-1 plans that we used above, let us compute which is more costly per 24-hour shift using the following calculation:

	$i = 7, f = 1/3$	$i = 9, f = 1/4$
Expected #units inspected per 24 hours	$24 \times AFI(0.1) = 24 \times 0.511$ =12.27	$24 \times AFI(0.1) = 24 \times 0.462$ =11.10
Expected #units produced per phase	$E_{Screening} = 10.91$ $E_{Sampling} = 30$	$E_{Screening} = 15.81$ $E_{Sampling} = 40$
Expected #phase swithing per 24 hours	$\frac{2 \times 24}{10.91+30} = 1.17$	$\frac{2 \times 24}{15.81+40} = 0.86$

For the plan i=7, f=1/3 the total cost per 24 hours is the cost of switching (\$50)(1.17) plus the cost of inspection (\$1)(12.27), totaling \$70.77. Using a similar calculation we reach the cost \$54.10 for the plan i=9, f=1/4. Hence the latter plan is less costly to the manufacturer.

$OC(p)$ Another common measure, which we encountered for lot-by-lot plans, is $OC(p)$. In the context of continuous sampling,

$OC(p)$ is defined as the long-run percent of accepted units during the sampling phase, given an incoming quality of p. The formula is given by

$$OC(p) = \frac{E_{Sampling}}{E_{Screening} + E_{Sampling}}. \qquad (6.4)$$

Non-Rectifying CSP-1

There exists an alternative to the rectifying CSP-1 scheme, where rather than rectifying or replacing non-conforming units, they are simply discarded. The choice between rectifying and discarding non-conforming items depends on costs, ease of rectifying or replacing items, and other related considerations. While operationally the non-rectifying scheme is similar to the rectifying scheme, the number of units exiting screening and sampling differ, and hence the resulting AOQ and AFI.

In the non-rectifying case, the average number of units exiting the screening phase and sampling phase is slightly lower than in the rectifying case, because every non-conforming unit encountered is simply discarded. The formulas for the non-rectifying case are given by

$$E_{Screening} = \frac{1 - (1-p)^i}{p(1-p)^{i-1}}$$

$$E_{Sampling} = \frac{1}{pf} - 1 \qquad (6.5)$$

The formulas for $OC(p)$ and for $AFI(p)$ remain the same as in equations 6.3-6.4. $AOQ(p)$ for the non-rectifying scheme is given by the formula:

$$AOQ(p) = \frac{p(1-p)^{i-1}}{(1-p)^{i-1} + \frac{f}{1-f}}. \qquad (6.6)$$

6.2   Continuous Sampling Plan 2 (CSP-2)

CSP-2 is a modification of CSP-1 designed to protect against returning to 100% inspection due to an isolated non-conforming unit. The plan achieves this by delaying the beginning of the screening phase. The CSP-2 plan is identical to CSP-1 except that switching from the sampling phase back to the screening (100% inspection) phase occurs if two non-conforming items are spaced less than i units apart. Note that the clearance number i is used in both screening and sampling phases. This procedure is depicted in Figure 6.4.

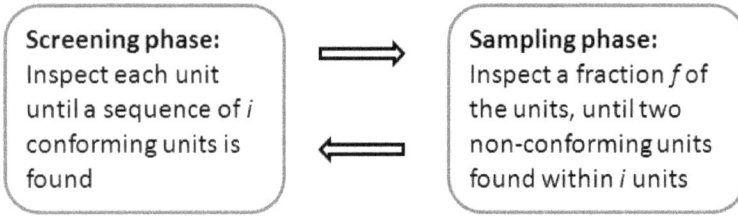

> **Screening phase:**
> Inspect each unit
> until a sequence of i
> conforming units is
> found

⟹
⟸

> **Sampling phase:**
> Inspect a fraction f of
> the units, until two
> non-conforming units
> found within i units

Figure 6.4: Schematic of the CSP-2 procedure

6.3   Mil-Std-1235 Tables

The standard that provides tables for continuous sampling plans is the Military Standard 1235, with the most recent version being Mil-Std-1235C (available from www.sqconline.com/download). This version is from 1988 and is now officially cancelled. However, continuous sampling plans are still in wide use and therefore the standard is often used to find sampling plans that yield a required AOQL.

Mil-Std-1235C contains tables for five types of continuous sampling schemes: CSP-1, CSP-2, CSP-F, CSP-V and the multi-level plan CSP-T.

CSP-1 and CSP-2

The CSP-1 table is shown in Figure ?6.5. To obtain a sampling plan we start with a required AOQL level (in the bottom row). For example, if we require AOQL=1%, then we must choose the column AOQL=0.79% which is the closest (marked in grey). There are multiple plans that correspond to this AOQL such as (f=1/2, i=36), (f=1/3, i=59), etc. Among these plans we can then choose according to minimum AFI or by minimizing costs. The same logic is used with the CSP-2 table (see Figure 6.5).

Figure 6.5: Mil-Std-1235C
Table for CSP-1 parameters

Values of i for CSP-1 Plans

Samp Freq Code Ltr	f	AQL in %															
		.010	.015	.025	.040	.065	0.10	0.15	0.25	0.40	0.65	1.0	1.5	2.3	4.0	6.5	10.0
A	1/2	1560	840	600	375	245	194	140	84	53	36	23	15	10	6	5	3
B	1/3	2550	1390	1000	620	405	321	232	140	87	59	38	25	16	10	7	5
C	1/4	3340	1820	1310	810	530	420	303	182	113	76	49	32	21	13	9	6
D	1/5	3960	2160	1550	965	630	498	360	217	135	91	58	38	25	15	11	7
E	1/7	4950	2700	1940	1205	790	623	450	270	168	113	73	47	31	18	13	8
F	1/10	6050	3300	2370	1470	965	762	550	335	207	138	89	57	38	22	16	10
G	1/15	7390	4030	2890	1800	1180	930	672	410	255	170	108	70	46	27	19	12
H	1/25	9110	4970	3570	2215	1450	1147	828	500	315	210	134	86	57	33	23	14
I	1/50	11730	6400	4590	2855	1870	1477	1067	640	400	270	175	110	72	42	29	18
J	1/100	16320	7810	5600	3465	2305	1820	1302	790	500	330	215	135	89	52	36	22
K	1/200	17420	9500	6810	4235	2760	2178	1583	950	590	400	255	165	106	62	43	26
		.018	.033	.046	.074	.113	.143	.198	0.33	0.53	0.79	1.22	1.90	2.90	4.94	7.12	11.46
		AOQL in %															

Values of i for CSP-2 Plans

Samp Freq Code Ltr	f	AQL* in %							
		0.40	0.65	1.0	1.5	2.5	4.0	6.5	10.0
A	1/2	80	54	35	23	15	9	7	4
B	1/3	128	86	55	36	24	14	10	7
C	1/4	162	109	70	45	30	18	12	8
D	1/5	190	127	81	52	35	20	14	9
E	1/7	230	155	99	64	42	25	17	11
F	1/10	275	185	118	76	50	29	20	13
G	1/15	330	220	140	90	59	35	24	15
H	1/25	395	265	170	109	71	42	29	18
I,J,K	1/50	490	330	210	134	88	52	36	22
		0.53	0.79	1.22	1.90	2.90	4.94	7.12	11.46
		AOQL in %							

Figure 6.6: Mil-Std-1235C
Table for CSP-2 parameters

Note: The tables in Mil-Std-1235 are based on *non-rectifying inspection*. However, it is easy to derive a rectifying sampling plan from the non-rectifying table by simply adding 1 to the clearance number i. For example, for AOQL=1% rectifying plans would have parameters (f=1/2, i=37), (f=1/3, i=60), and so on.

At the top of the CSP-1 table we see AQL listed. This was an attempt to make the tables similar to those in Mil-Std-105 and 414. However, the AQL is meaningless and should not be used for any purpose.

The code letters in the left column are used for further limiting the relevant plans based on the rate of production. For slow production, very large values of i are not practical. The table in Figure 6.7 gives the range of code letters that are suitable for different production rates. *Production interval* means a meaningful interval such as a shift.

Plan CSP-F for Short Production Runs

Mil-Std1235 includes a plan called CSP-F, which is nearly identical to CSP-1 except that it has smaller clearance numbers to

SAMPLING FREQUENCY CODE LETTERS

Number of Units in Production Interval	Permissible Code Letters
2–8	A,B
9–25	A through C
26–90	A through D
91–500	A through E
501–1200	A through F
1201–3200	A through G
3201–10,000	A through H
10,001–35,000	A through I
35,001–150,000	A through J
150,001–up	A through K

Figure 6.7: Mil-Std-1235 code letter table, used for choosing plans that are appropriate for the production rate

allow for short production runs or time-consuming inspection (to avoid bottlenecks). The idea is to apply this plan to a predetermined number of units N. An example of a CSP-F table is shown in Figure 6.8 for a particular AOQL.

Multi-level Plan CSP-T

Mil-Std-1235 includes a multi-level plan called CSP-T. In this plan there are multiple sampling phases which have different sampling ratios. The idea is that once in sampling phase, there is a provision for further reducing the amount of inspection if the quality is found to be high.

Plan CSP-V

The final plan, called CSP-V, includes a single screening and sampling stage. If a non-conforming item is found in the sampling phase, screening is resumed, but with a lower clearance number x. This is an alternative to reducing the sampling fraction f for cases where operationally it does not make sense to further reduce the sampling fraction. The CSP-V scheme and table include 3 parameters for a given AOQL: in addition to i and f, there is the lower clearance number x (see Figure 6.10).

Values of i for CSP-F Plans

AQL* - .010%
AOQL - .018%

Samp Freq Code Ltr	A	B	C	D	E	F	G	H
N	1/2	1/3	1/4	1/5	1/7	1/10	1/15	1/25
1-500	347	376	387	392	398	402	405	407
501-600	400	432	449	458	461	464	470	472
601-700	441	485	502	517	519	523	529	533
701-800	482	530	577	585	589	591	594	596
801-1,000	545	618	647	662	678	689	697	703
1,001-1,500	679	799	843	870	900	903	920	935
1,501-2,000	784	942	1008	1044	1082	1108	1128	1143
2,001-3,000	929	1163	1264	1320	1380	1423	1455	1479
3,001-4,000	1029	1328	1462	1538	1620	1679	1723	1757
4,001-5,000	1101	1458	1624	1718	1822	1896	1952	1996
5,001-6,000	1156	1564	1759	1871	1996	2086	2154	2208
6,001-7,000	1199	1651	1874	2004	2149	2255	2335	2398
7,001-8,000	1234	1725	1974	2125	2285	2407	2499	2572
8,001-9,000	1262	1789	2061	2224	2408	2545	2649	2732
9,001-10,000	1286	1844	2138	2317	2520	2671	2788	2880
10,001-11,000	1306	1891	2207	2400	2622	2788	2917	3018
11,001-12,000	1323	1933	2269	2496	2716	2897	3037	3148
12,001-15,000	1363	2034	2420	2666	2957	3181	3356	3497
15,001-20,000	1405	2146	2598	2898	3265	3554	3787	3975
20,001-30,000	1449	2271	2808	3183	3670	4076	4414	4698

Figure 6.8: Mil-Std-1235 table for CSP-F, for AOQL=0.018%

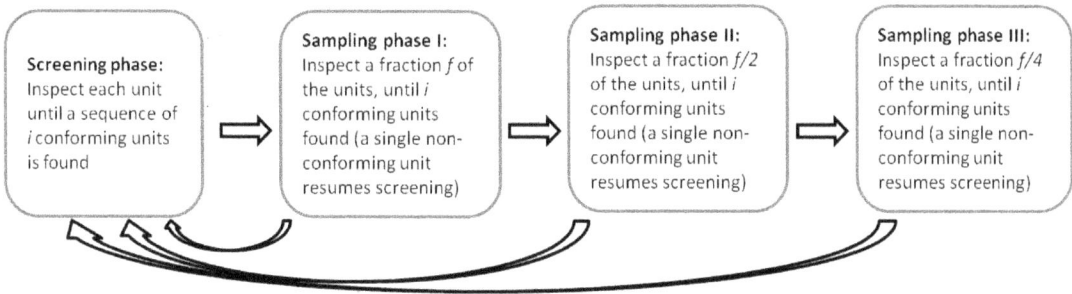

Figure 6.9: Mil-Std-1235 CSP-T multi-level plan

6.4   Online Plans

In place of using the Mil-Std-1235 tables for obtaining CSP-1, we can use the application on SQCOnline.com. Currently, the website offers only CSP-1 plans using the Mil-Std-table. The input screen requests the user for an AOQL level. The output screen includes all the pairs i and f for rectifying as well as non-rectifying inspection (see Figure 6.11).

6.5   Skip-Lot Sampling Plans

Skip-lot sampling is the application of continuous sampling to batches rather than units. The purpose is to reduce the amount of inspection when the quality of recent batches is good and the supplier has demonstrated a proven track record. Skip-lot sampling is only applicable when the items arrive from the same source. They are especially useful in laboratory testing, where continuous batches arrive.

Two common skip-lot procedures are SkSP-1 and SkSP-2. The inspection procedure for SkSP-1 is as follows:

Base sampling plan: Determine an inspection plan (this can be any sampling plan for attributes or variables; based on a single attribute or multiple attributes).

Screening phase: apply the sampling plan to each batch until finding i consecutive accepted batches, then move to sampling phase.

Values of i and x for CSP-V Plans

Samp Freq Code Ltr	f	0.40	0.65	1.0	1.5	2.5	4.0	6.5	10.0	
A	1/2	60	39	27	18	12	9	6	3	i
		20	13	9	6	4	3	2	1	x
B	1/3	96	63	42	27	18	12	9	6	i
		32	21	14	9	6	4	3	2	x
C	1/4	120	81	54	36	24	15	12	6	i
		40	27	18	12	8	5	4	2	x
D	1/5	144	96	63	42	27	18	12	9	i
		48	32	21	14	9	6	4	3	x
E	1/7	177	120	78	51	33	21	15	9	i
		59	40	26	17	11	7	5	3	x
F	1/10	213	144	93	60	39	24	18	12	i
		71	48	31	20	13	8	6	4	x
G	1/15	258	174	114	72	48	30	21	12	i
		86	58	38	24	16	10	7	4	x
H	1/25	318	213	138	90	60	36	24	15	i
		106	71	46	30	20	12	8	5	x
I	1/50	405	273	177	114	75	45	30	21	i
		135	91	59	38	25	15	10	7	x
J	1/100	498	333	216	138	90	54	39	24	i
		166	111	72	46	30	18	13	8	x
K	1/200	594	399	258	165	108	63	45	27	i
		198	133	86	55	36	21	15	9	x
AOQL in %		0.53	0.79	1.22	1.90	2.90	4.94	7.12	11.46	

AQL* in %

AQL in %

Figure 6.10: Mil-Std-1235 plan CSP-V

Continuous Sampling

Military Standard 1235C: Continuous Sampling Plan 1 (CSP-1)

This application gives the parameters needed for applying Military Standard 1235C (CSP-1), for a given AOQL.

More about acceptance sampling plans

Enter the required AOQL:

Average Outgoing Quality Limit (AOQL): [1.22%] [▼] The average percent of non-conforming items in your process, after applying the CSP-1 procedure (AOQL)

[Submit]

Continuous Sampling

Military Standard 1235C: Continuous Sampling Plan 1 (CSP-1)

Below are the different pairs of i and f that meet your AOQL requirement. The decision which pair to use should be guided by practical matters (costs of sampling, costs of switching, production speed, etc.)

CSP-1 alternates between screening and sampling phases. Start with screening (100% inspection) until finding i consecutive conforming items. Then, move to sampling: inspect a fraction f of items until a single non-conforming item is found. In that case, return to screening.

Sampling fraction (f)	Clearance number (i) for non-rectifying inspection	Clearance Number (i) for rectifying inspection
0.500	23	24
0.333	38	39
0.250	49	50
0.200	58	59
0.143	73	74
0.100	89	90
0.067	108	109
0.040	134	135
0.020	175	176
0.010	215	216
0.005	255	256

Figure 6.11: Input and output screens on SQCOnline.com for CSP-1

Skip-lot phase: apply the sampling plan only to a fraction f of the batches. If a batch is rejected, return to the screening phase.

Recall the *probability of acceptance OC(p)*, which is defined as the proportion of accepted batches of quality p. In skip-lot sampling, $OC(p)$ will increase as i and f decrease.

The *Average Outgoing Quality* (AOQ) in skip-lot sampling is defined as the long-term proportion of accepted lots.

In terms of the amount of expected inspection, skip-lot sampling leads to less inspection.

The SkSP-2 scheme is more specific than SkSP-1 in that the base sampling plan applied to a batch, called a *reference sampling plan*, is based on using sampling plans for attributes from ANSI/ASQC Z.4. This yields formulas for quantities such as AOQ and ASN. For example, the average number of items inspected, or the *Average Sample Number*(ASN) of SkSP-2 for a given quality p is given by

$$ASN_{SkSP}(p) = \frac{f}{f + (1 - f)(OC(p))^i} \times ASN(p), \qquad (6.7)$$

where ASN(p) is the expected number of inspected items using the reference sampling plan, and $OC(p)$ is the probability of accepting the batch using the reference sampling plan.

Procedures for skip-lot sampling are given in the ISO 2859 standard, Part 3. The terminology used in the standard is different: *qualification period* means the screening phase; *skip-lot interruption* means the reversion to 100% inspection. The procedures there are more involved and are based on using sampling plans for attributes from ISO 2859 Part 1 (ANSI/ASQC Z1.4). In general, sampling fractions of f=1/2, 1/3, 1/4, and 1/5 with clearance numbers in the range i=10-20 are used. The standard also includes several additional requirements.

One point to note is that according to the ISO standard skip-lot sampling may be used instead of reduced inspection if it is more cost effective. The standard even advocates using skip-lot sampling over reduced sampling because "it gives the producer a greater incentive to aim for and maintain a better quality level."

6.6   Problems

1. *Under See* produces optical goggles on demand. Users enter their prescription information as well as other choices (such as tint) into the company's website. *Under See* would like to employ continuous sampling to audit their quality before shipping a product to the consumer. They want to assure an average outgoing quality of no worse than 2%.

   (a) What is the advantage of continuous sampling over lot-by-lot sampling in this case?

   (b) Find two CSP-1 rectifying plans with the highest possible sampling fractions. Report the plan parameters and explain the procedure and decision rules to employ each plan.

   (c) Plot the AOQ curves for the two plans that you found in (b), overlaying the two curves on the same plot. Compare the curves in terms of the expected outgoing quality (AOQ) for different incoming quality levels (p).

   (d) Compute the expected number of units inspected under screening and under sampling for each of the plans, if the incoming quality is 1%.

   (e) Compute $AFI(0.01)$ for each of the plans. Which plan is expected to inspect less pairs of goggles?

   (f) If the cost of switching from screening to sampling and vice versa is EUR 2 and the cost of inspecting a unit is EUR .50, what is the total cost of inspection per 1000 produced items if the incoming quality is 1%?

2. *Save Trees* is a vendor of recycled paper. It receives batches of 2500 pages of recycled paper from a recycling facility every week. Each sheet of paper can either be conforming or non-conforming. To assure that the quality of the recycled paper is sufficient, *Save Trees* would like to start deploying acceptance sampling, using an ISO sampling plan. The quality requirement is AQL=0.25%.

   (a) Find the ANSI/ASQC Z1.4 single-stage sampling plan, using normal, Type II inspection.

(b) After several years of inspection where nearly all batches
have been accepted, *Save Trees* would like to use skip-lot
sampling to reduce inspection costs. Find a non-rectifying
skip-lot scheme that will guarantee a long-term percent
of accepted batches of 0.2%, and which has the largest
possible sampling fraction. Report the plan parameters and
describe how the skip-lot inspection should be carried out.

(c) What is the expected fraction of inspected batches if the
incoming quality is p=0.2%?

(d) Find the ANSI/ASQC Z1.4 single-stage sampling plan,
using reduced, Type II inspection. Compute $OC(0.002)$.

(e) If the incoming quality remains stable at p=0.2% and the
cost of inspecting one page is Yen 1, is skip-lot sampling
cost-effective compared to reduced sampling? (Assume that
all batches are inspected using reduced sampling). Show
the cost for a stream of 1,000 batches.

Index

www.ingramcontent.com/pod-product-compliance
Lightning Source LLC
Chambersburg PA
CBHW082106210326
41599CB00033B/6611